Second Nature

Second Nature

BRAIN SCIENCE AND HUMAN KNOWLEDGE

Gerald M. Edelman

YALE UNIVERSITY PRESS
New Haven and London

Published with assistance from the Louis Stern Memorial Fund.

Designed by Sonia L. Shannon.

Set in Stempel Schneidler type by Integrated Publishing Solutions.

Printed in the United States of America.

Library of Congress Cataloging-in-Publication Data
Edelman, Gerald M.
Second nature : brain science and human knowledge / Gerald M.
Edelman.
p.   cm.
Includes bibliographical references and index.
ISBN-13: 978-0-300-12039-4 (clothbound : alk. paper)
ISBN-10: 0-300-12039-7 (clothbound : alk. paper)
1. Brain. 2. Neurosciences. I. Title.
QP376 E323 2006
612.8'2—dc22   2006007376

A catalogue record for this book is available from the British Library.

The paper in this book meets the guidelines for permanence and
durability of the Committee on Production Guidelines for Book
Longevity of the Council on Library Resources.

10 9 8 7 6 5 4 3 2 1

For Judith, Eric, and David

And going on, we come to things like evil, and beauty, and hope . . .

Which end is nearer to God; if I may use a religious metaphor. Beauty and hope, or the fundamental laws? I think that the right way, of course, is to say that what we have to look at is the whole structural interconnection of the thing: and that all the sciences, and not just the sciences but all the efforts of intellectual kinds, are an endeavor to see the connections of the hierarchies, to connect beauty to history, to connect history to man's psychology, man's psychology to the working of the brain, the brain to the neural impulse, the neural impulse to the chemistry, and so forth, up and down, both ways. And today we cannot, and it is no use making believe that we can, draw carefully a line all the way from one end of this thing to the other, because we have only just begun to see that there is this relative hierarchy.

And I do not think either end is nearer to God.

—Richard Feynman

Contents

Preface

THIS BOOK WAS PROMPTED by my efforts, desultory and otherwise, to understand how progress in brain science bears on issues of human knowledge. The results of my thoughts on these issues are couched in more lenient and heterogeneous terms than those of philosophers dedicated to traditional epistemology. I consider this difference to be a useful starting point for further explorations of how we know.

A casual glance at the table of contents will reveal that I also consider that the understanding of consciousness is critical to the enterprise. With this in mind, here is how I propose to proceed:

First, I will argue that a number of important consequences ensue if we show how consciousness is based in brain action. In doing so, I shall *assume* that we do understand this basis, and I will lay out the implications of such an understanding. I will then describe some of the essential features of the brain and the concepts necessary to understanding how it works. With this description in hand, we can address the nature of consciousness itself. We will then be in a position to revisit the consequences for science and human knowledge of our understanding of the grounds of consciousness.

In realizing this project, I intend to avoid technical details. I have covered these extensively in other books and papers. Instead, in describing the brain, I shall rely as much as possible on concrete examples and metaphors.

I urge the reader to look on this work as an initial exploratory effort designed to prompt new thoughts on how we come to know the world and ourselves. Many gaps remain and much more must be accomplished in both neuroscience and psychology before a comprehensive picture of thought and knowledge can be glimpsed. Consider what follows to be a few first strokes.

Acknowledgments

I AM GRATEFUL TO KATHRYN Crossin, Bruce Cunningham, Joseph Gally, Ralph Greenspan, and George Reeke for their close reading of the text and their critical suggestions. I also thank Diana Stotts for her excellent and patient help in preparing the manuscript. Colleagues at The Neurosciences Institute were a source of many useful exchanges during the writing of this book.

Introduction

FROM TIME TO TIME I HAVE a dream. In it, the historian Henry Adams appears, moaning about complexity, muttering about the Virgin and the Dynamo. There usually is no more to the dream than that. When I remember enough detail in my waking state, I connect it to the famous chapter in *The Education of Henry Adams*.[1] In that chapter, Adams recounts his sense of inadequacy before the forty-foot dynamos that his friend Langley the engineer showed him at the Paris Exhibition of 1900. Adams contrasts the complexity of these engines to the simplicity of the religious turn to the Virgin Mary. That theme, its variations, and Adams's sense of not being comfortable in his time run through the *Education*.

Adams, a scion of the great family descended from John Adams, was an accomplished historian. His sense of alienation prompts speculation. Could it have simply been the symptom of clinical depression? Could it have been entwined with the circumstances leading to his wife's suicide? Or could it have reflected a genuine rift between the way a person sees the world from the standpoint of science and the way that person sees it reflected in the humanities?

We do not know. But one thing is sure. There is a divorce between science and the humanities, and between the so-called hard sciences such as physics and the human sciences such as sociology. Perhaps my recurrent dreams of Henry Adams come from my persistent interest in the origin of this estrangement.

I have long puzzled over the gap between scientific explanations and everyday experience, whether by individuals or in historical settings. Is the divorce between science and the humanities inevitable? Can the human sciences be reconciled with the hard sciences?

Views on these questions have ranged widely, and some might say that they aren't worth bothering with. As this book attests, I believe the opposite—that understanding how we arrive at knowledge, whether by scientific inquiry, by reason, or by happenstance, is of major importance. Wrongheadedness, severe reductionism, or insouciance can each have unfortunate long-range consequences for human welfare.

This book is the result of a line of thought leading to what I have called brain-based epistemology. This term refers to efforts to ground the theory of knowledge in an understanding of how the brain works. It is an extension of the notion of naturalized epistemology, a proposal made by the philosopher Willard Van Orman Quine.[2]

My line of argument differs from his, which stopped, as it were, at the skin and other sensory receptors. I deal with the issue by considering a wider-ranging interaction—that between brain, body, and environment. I believe that above all, it is particularly important to understand the basis of consciousness. Quine with his usual ironic candor said,

> I have been accused of denying consciousness, but I am not conscious of having done so. Consciousness is to me a mystery, and not one to be dismissed. We know what it is like to be conscious, but not how to put it into satisfactory scientific terms. Whatever it precisely may be, consciousness is a state of the body, a state of nerves.
>
> The line I am urging as today's conventional wisdom is not a denial of consciousness. It is often called, with more reason, a repudiation of mind. It is called a repudiation of mind as a second substance, over and above body. It can be described less harshly as an identification of mind with some of the faculties, states, and activities of the body. Mental states and events are a special subclass of the states and events of the human or animal body.[3]

I believe we are now in a position to reduce the mystery. In this book, I lay out thoughts that reveal this position and bear directly on how we know, on how we discover and create, and on our search for truth. I follow in the footsteps of William James, who pointed out that consciousness is a process whose function is knowing.[4]

There is nature and human nature. How do they intersect? The title I have chosen reflects this question and is to some extent a play on words. The term "second nature" usually refers to an act done spontaneously, easily and without the need for exertion or learning. I use the term here to include this meaning but also to call attention to the fact that our thoughts often float free of our realistic descriptions of nature. They are a "second nature." I aim to explore here how nature and second nature interact.

one
The Galilean Arc and Darwin's Program

Almost everything that distinguishes the modern world
from earlier centuries is attributable to science, which
achieved its most spectacular triumphs in the
seventeenth century.

— BERTRAND RUSSELL

The Origin of Species *introduced a mode of thinking*
that in the end was bound to transform the logic of
knowledge, and hence the treatment of morality,
politics, and religion.

— JOHN DEWEY

Something definite happens when to a certain brain
state a certain 'sciousness corresponds.

— WILLIAM JAMES

HENRY ADAMS DIDN'T KNOW the half of what was coming. But he did sense a transformation of our existence by scientific technology. We are in the midst of a revolution: communication, computers, the Internet, the explosion of travel by land and air, atomic power, biological manipulation of our genetic makeup. One could go on and on about the technological substrate and the globalization that has changed the pace of our lives, the modes of our thought, our place in nature, and our threat to it.

What has happened to our conception of nature and of our second nature? To answer this question we have to take a longer view of Western science, particularly of physics and biology. I pick two figures, Galileo Galilei and Charles Darwin, to highlight the developments that have so changed our lives.

First, Galileo, who can be taken to represent the birth in the seventeenth century of modern physics, the broadest of modern sciences. The philosopher Alfred North Whitehead, in his book *Science and the Modern World,* called Galileo's achievement "the quiet commencement of the most intimate change in outlook which the human race had yet encountered."[1] Surely, we must be impressed by the arc of modern physics ranging from Galileo's ideas on the heavens and his experiments on inertia to our present cosmology and theories of matter. We must confront the weird domain of the very small described by quantum mechanics as well as the grand elegance of general relativity as it opens up vistas of the very large, the universe itself. So now the Galilean arc ranges from nuclear power to solid-state physics, to the exploration of space, and to the origin of the universe itself in the Big Bang.

Even before these advances, a vision of the basis of life, the evolution of living things, was laid down by Charles Darwin in the second half of the nineteenth century.[2] Darwin's development of the idea of natural selection provided the theoretical basis for understanding life itself, particularly when it was coupled with Mendelian genetics in the twentieth century.[3] The further development of molecular biology in the latter part of that century has made it possible to change the very basis of biological reproduction.

In looking over the domains of nature it may appear that, if we include Darwin, the Galilean arc has provided an enlightened understanding of all major subject categories: galaxies, stars, and planets, the structure of matter, the nature of genes and biological evolution. Henry Adams's plate today would be more than full of scientific matters covering much of our existence. But there is a gap or an incompleteness in the Galilean arc. We have not yet scientifically founded the bases of consciousness in the brain, an issue that, until recently, has been left to the philosophers.

There are reasons for this. Until recently, noninvasive methods of examining events in the brain were lacking. More than that, consciousness is a first-person affair, whereas the objective methodology of science is a third-person affair. Opinions, subjectivity, and the like cannot be admitted in scientific experiments. An equally large factor affecting the scientific approach to consciousness can be attributed to the influential thoughts of René Descartes.[4] Sometime after Galileo, Descartes essentially removed the mind from nature. He did this by thought alone, concluding that there were two substances: *res extensa,* extended things that were susceptible to physics, and *res cogitans,* thinking things that were not extended in space and that were unavailable to physics. Descartes's dualism and its various subsequent derivations have had a profound effect on the approach to consciousness as a valid scientific target.

This state of affairs is most curious. In principle, no subject is a priori immune from scientific inquiry. Yet the very ground of our awareness has been left outside the pale! Science is imagination in the service of the verifiable truth. And as such, imagination is actually dependent on consciousness. Science itself is so dependent. As the great physicist Erwin Schrödinger observed, no scientific theory in physics includes sensations and perceptions and to get ahead it must therefore assume these phenomena as being outside of science's grasp.[5]

Must we accept this state of affairs? Or can science complete the Galilean arc? If it cannot, must it leave the ground of consciousness to the philosophers, to the humanities, and thereby acquiesce to the divorce that so concerned Henry Adams?

Thanks to discoveries about the brain and advances in brain theory in the past twenty years, it seems we do not have to remain in this predicament. We can study consciousness even in the face of subjectivity. My aim here is to show how. But first, let us turn to the significance of a scientific understanding of consciousness.

I have found that some people do not believe that a scientific account of consciousness offers much in the way of consequences. My remarks here are not specifically aimed at these doubters, but I hope they will persuade some at least to consider the contrary position. I start with a big assumption: that we have a satisfactory scientific theory of consciousness based on brain activity. What would its significance be?

First, it would clarify the relation between mental and physical events and clear up some outstanding philosophical puzzles. We would no longer have to consider dualism, panpsychism, mysterianism, and spooky forces as worth pursuing.[6] Time would be saved, at the least. And in clarifying these issues, we would have a better view of our place in the natural order. We would be able to corroborate Darwin's view that the human mind is the outcome of natural selection and thereby complete his program.[7]

We would also have a better picture of the bases of human illusions, useful and otherwise. One illusion I hope to dispel is the notion that our brains are computers and that consciousness could emerge from computation. Furthermore, a successful theory of consciousness might clarify the place of values in a world of facts. In connection with both of these issues, a brain-based theory would be of great use in understanding psychiatric and neuropsychological syndromes and diseases.

Tangent to these matters, a brain-based theory might contribute to our notions of creativity. It might even provide a clearer view of the connection of objective descriptions derived from hard science to normative issues that arise in aesthetics and ethics. To that degree, it may help undo the divorce between science and the humanities.

Above all, achieving these ends may contribute to and affect the formulation of a biologically based epistemology—an account of knowledge that relates truth to opinion and belief, and thought to emotion by including aspects of brain-based subjectivity in an analysis of human knowledge.

The most remarkable outcome of a satisfactory brain theory would be the construction of a conscious artifact.[8] Although that goal is presently in the realm of fantasy, scientists at The Neurosciences Institute in La Jolla, California, have already built brain-based devices that have perceptual and memorial capabilities. Of course, a minimum requirement for us to believe that we had constructed a conscious device would rest in its ability to report, through a language, its internal phenomenal states while we measure its neural and bodily performance. This requirement is presently far from being met. But if it were, we would have an unparalleled opportunity to explore brain, body, and environment as they interact in such a device. Would it "see" or "sense" the world in ways we cannot imagine? Only a receipt of messages from outer space would exceed this enterprise in excitement. We shall have to wait.

I propose now to provide some support for the assumption I made that we have a satisfactory theory of consciousness. I shall do so by presenting a brief account of consciousness and of the brain dynamics from which it emerges. Following that, we can return to analyze their consequences in greater detail.

two

Consciousness, Body, and Brain

*Like the entomologist in search of brightly colored
butterflies, my attention hunted, in the garden of gray
matter, cells with delicate and elegant forms, the
mysterious butterflies of the soul.*

— SANTIAGO RAMÓN Y CAJAL

I HAVE WRITTEN EXTENSIVELY on the details of brain structure and dynamics as they relate to perception, memory, and
consciousness. I have no intention of repeating these details
here. Instead, I shall describe some of the main features of
consciousness. Then I will give a brief account of brain activity in terms of a theory called Neural Darwinism.[1] This will
allow me to show how consciousness emerges from brain dynamics. I shall not hesitate to make large statements without
detailed proof; such proof can be found elsewhere.[2]

We all know implicitly what consciousness is. It is what
you lose on entering a dreamless deep sleep and, less com-
monly, deep anesthesia or coma. And it is what you regain
after emerging from these states. In the awake conscious state,
you experience a unitary scene composed variably of sensory
responses—sight, sound, smell, and so on—as well as images,
memories, feeling tones and emotions, a sense of willing or
agency, a feeling of situatedness, and other aspects of awareness. Being conscious is a unitary experience in the sense that
you cannot at any time become totally aware of just one thing
to the complete exclusion of others. But you can direct your
attention to various aspects of a less inclusive but still unitary
scene. Within a short time, that scene will vary in one degree
or another and, though still integrated, will become differentiated, yielding a new scene. The extraordinary fact is that the
number of such privately experienced scenes is apparently limit-

less. The transitions seem to be continuous, and in their complete detail they are private, first-person subjective experiences.

Conscious states are often, but not always, about things or events, a property called intentionality. But they do not necessarily always show this property; they can, for example, be about a mood. There is often a just-aware "fringe," as William James called certain barely perceived states. Conscious states can also involve the awareness of agency or the willing of an action.

The property most often described as particularly mysterious is the phenomenal aspect of consciousness, the experience of qualia. Qualia are, for example, the greenness of green and the warmness of warmth. But several students of the subject, myself included, go beyond these simple qualities and consider the whole ensemble of conscious scenes or experiences to be qualia.

Many consider explaining qualia to be the acid test of a consciousness theory. How can we explain not only qualia but all the other features of consciousness? The answer I propose is to look into how the brain works, formulating a global brain theory that can be extended to explain consciousness. Before I do so, however, one more distinction will prove useful. As human beings, we know what it is like to be conscious. Moreover, we are conscious of being conscious and can report on our experience. Although we cannot experience the consciousness of members of another species, we surmise that animals like dogs are conscious. We do this on the basis of their behavior and the close similarity of their brains to ours. But we do not usually attribute consciousness of consciousness to them.

This is the basis for a useful distinction. Dogs and other mammals, if they are aware, have primary consciousness. This is the experience of a unitary scene in a time period of at the most seconds that I call the remembered present—a bit like the illumination by a flashlight beam in a dark room. Although they are aware of ongoing events, animals with primary consciousness are not conscious of being conscious and do not have a concept of the past, the future, or a nameable self.

Such notions require the ability to experience higher-order consciousness, and this depends on having semantic or symbolic capabilities. Chimpanzees appear to have the rudiments of these capabilities. In our case, they exist in full flower because we have syntax and true language. With the ability to speak, we can free ourselves temporarily from the limitations of the remembered present. Nonetheless, at all times when higher-order consciousness is present we also possess primary consciousness.

Against this background, let us turn to the organ responsible for all these extraordinary traits: the brain. The human brain weighs about three pounds. It is one of the most complicated material objects in the known universe. Its connectivity is awe-inspiring: the wrinkled cortical mantle of the brain

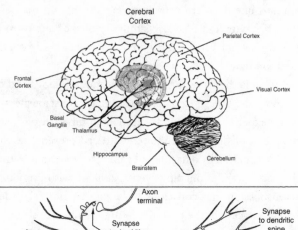

Cerebral
Cortex

Parietal Cortex

Frontal
Cortex

Visual Cortex

Basal
Ganglia

Thalamus

Hippocampus

Cerebellum

Brainstem

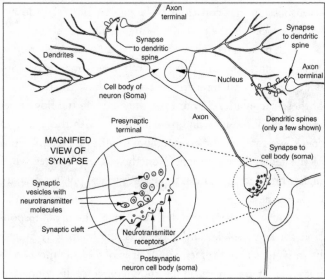

Axon
terminal

Synapse
to dendritic
spine

Synapse
to dendritic
spine

Dendrites

Axon
terminal

Cell body of
neuron (Soma)

Nucleus

Dendritic spines
(only a few shown)

Presynaptic
terminal

Axon

Synapse to
cell body (soma)

MAGNIFIED
VIEW OF
SYNAPSE

Synaptic
vesicles with
neurotransmitter
molecules

Synaptic cleft

Neurotransmitter
receptors

Postsynaptic
neuron cell body (soma)

Figure 1.

Top: Relative locations of major parts of the human brain. The cerebral cortical mantle, which has about thirty billion neurons, receives projections from the thalamus and sends reciprocal projections back; this constitutes the thalamocortical system. Beneath the mantle are three major cortical appendages: the basal ganglia and cerebellum (both of which regulate movement), and the hippocampus, which is necessary for memory. Below them is the oldest part of the brain in evolutionary terms, the brain stem, which contains several diffusely projecting value systems.

Bottom: Synaptic connections between two neurons. An action potential traveling down the axon of the presynaptic neuron causes the release of a neurotransmitter into the synaptic cleft. The transmitter molecules bind to receptors in the postsynaptic membrane, changing the probability that the postsynaptic cell will, in turn, fire its own action potential. Particular sequences of activity can either strengthen or weaken the synapse, changing its efficacy. (Because of the number of different shapes and kinds of neurons, this drawing is a greatly simplified cartoon.)

(figure 1, *top*) has about thirty billion nerve cells or neurons and one million billion connections. The number of possible active pathways of such a structure far exceeds the number of elementary particles in the known universe.

This is not the place to go into detail about how the brain gives rise to consciousness. I have done that in several books, which may be consulted. But I do want to provide a working picture of brain structure and activity. I propose to use a mixture of down-to-earth description, analogy, and metaphor—just enough to give an idea of how consciousness arises.

To start with, let's consider the fundamental cells that carry signals in the brain. These are the neurons, which have a treelike set of branches (dendrites) and usually a single extended process (the axon) that serves to connect one neuron to another. This connection, called the synapse (figure 1, *bottom*), is a critical element in ensuring the function of brain circuits. This is so because electricity traveling down the axon releases little packets of chemicals called neurotransmitters at the synapse. These chemicals cross the small distance inside the synapse and bind to certain receptors present often at the dendrites of the receiving cell. If the release happens often enough, the receiving or postsynaptic cell fires and can repeat the process and signal yet another cell. Imagine such a process summing up across a myriad of synapses, and you will get an idea of why with modern methods we can actually record the otherwise minute currents and potentials over the scalp. Neurophysiologists can in fact record more precisely from single cells by invading the brain and inserting microscopic electrodes within individual neurons.

A key property of synapses is that they are plastic: various activities and biochemical events can change their strength. These changes can in turn determine which neuronal pathways are selected to transmit signals. Patterns of such changes in synaptic strength provide a basis for memory. At this point, it may be useful to mention that synapses come in two flavors: excitatory and inhibitory. Both can exhibit plasticity; together they help select the functioning signal pathways of the brain.

Now an important next step in this bowdlerized account is to point out that the overall anatomical connections and pathways in the brain of a given animal species are selected during evolution and development. The result is a stunning set of different brain areas and cell collections called nuclei. Each of these has both short-range and long-range inputs and outputs.

Let us look at the visual pathway in monkeys as an example. Light, striking cells in the retina, excites the optic nerve, whose signals ultimately reach a structure called the thalamus, a central player in our story. The thalamus is a small structure that is of great importance in any account of consciousness. Thalamic neurons mediating vision send axons to an area of the cerebral cortex called V_1. From there, all kinds of pathways within the cortex are elaborated to areas called V_2, V_3, and V_4, among others. Indeed, at least thirty-three different cortical areas are involved in one way or another in the process of vision.

Two important facts about this and several other sensory systems have emerged. The first is that, in general, each brain area is functionally segregated: in vision, V_1 for orientation of objects, V_4 for color, V_5 for object motion. The second fact is that there is no one area controlling and coordinating the responses of all the rest when a complex visual signal comes from, say, a colored moving object of particular shape. As we shall see, the brain nevertheless has means to coordinate the segregated perceptual events that occur when such a stimulus strikes the retina. The net result of such coordination is perceptual categorization—the carving up of the world of inputs into objects significant for a given animal species' recognition. The brain carries out pattern recognition. We could go on about sensory systems other than vision, but the principles are similar even if their receptors and inputs differ.

What about outputs? Well, different sensory areas connect to "higher" areas in the cortex so that the brain speaks mainly to itself. Of course, one set of cortical areas sends motor output signals to the spinal cord and thence to our muscles to elicit various actions and movements. Furthermore, the cortex receives additional inputs from, and yields outputs to, a number of subcortical structures besides the thalamus. These (see figure 1) include the basal ganglia and cerebellum, which help to regulate movement, and the hippocampus, which helps establish long-term memory of events and episodes by interacting with the cortex.

So far, what I have said could superficially be thought to describe a system analogous to an electronic device such as a computer. Indeed, in many scientific circles, there remains a widespread belief that the brain is a computer. This belief is mistaken for a number of reasons.[3] First, the computer works by using logic and arithmetic in very short intervals regulated by a clock. As we shall see, the brain does not operate by logical rules. To function, a computer must receive unambiguous input signals. But signals to various sensory receptors of the brain are not so organized; the world (which is not carved beforehand into prescribed categories) is not a piece of coded tape. Second, the brain order that I have briefly described is enormously variable at its finest levels. As neural currents develop, variant individual experiences leave imprints such that no two brains are identical, even those of identical twins. This is so in large measure because, during the development and establishment of neuroanatomy, neurons that fire together wire together. Furthermore, there is no evidence for a computer program consisting of effective procedures that would control a brain's input, output, and behavior. Artificial intelligence doesn't work in real brains. There is no logic and no precise clock governing the outputs of our brains no matter how regular they may appear.

Last, it should be stressed that we are not born with enough genes to specify the synaptic complexity of higher brains like ours. Of course, the fact that we have human brains and not chimpanzee brains does depend on our gene networks. But these gene networks, like those in the brain themselves, are enormously variable since their various expression patterns depend on environmental context and individual experience.

If the mammalian brain is not a computer, what is it? How does it work? We must answer these questions before we can explain the brain bases of consciousness.

three
Selectionism

A PREREQUISITE FOR CONSCIOUSNESS

Theories have four stages of acceptance: i) this is worthless nonsense; ii) this is an interesting, but perverse, point of view; iii) this is true, but quite unimportant; iv) I always said so.

—J. B. S. HALDANE

THE DESCRIPTIONS I HAVE given of consciousness and the brain now have to be connected in a satisfactory way. This will require the presentation of a theory that accounts for coherent brain action in the absence of computation. It will also entail the exploration of a number of essential concepts that are likely to be unfamiliar. To make them understandable, I am going to use a number of biological examples and some non-biological analogies. I will then connect them to our main task: to see how consciousness evolved and how it arises in individual brains.

Before we turn to theoretical issues, we must not lose sight of one set of facts: The brain is embodied and the body is embedded. First, consider embodiment. All of the activities I described in the last chapter depend on signals to the brain from the body and from the brain to the body. The brain's maps and connections are altered not only by what you sense but by how you move. In turn, the brain regulates fundamental biological functions of your body's organs in addition to controlling the motions and actions that guide your senses. These functions include fundamental aspects of sex, breathing, heartbeat, and so on, as well as the responses that accompany emotion. If we include the brain as your favorite organ, you *are* your body.

Second, consider embeddedness. Your body is embedded and situated in a particular environment, influencing it and being influenced by it. This set of interactions defines your econiche, as it is called. It is well to remember that the human species evolved (along with the brain) in a sequence of such niches. I emphasize these facts because, for brevity, I will often talk of the brain without reference to the other two members of the critical triad, the body and the econiche. Remember, when that happens, that the critical triad is still at the back of my mind.

Now to the theory to provide a basis for understanding consciousness. Such a theory must account for both the diversity and the regularity of brain responses in the absence of the control by logic and a precise clock that are the hallmarks of a computer.

Where can we turn after relinquishing the notion of computation? The answer is provided by turning to Darwin's fundamental idea of population thinking.[1] Darwin proposed that categories (of characters or of species) could arise by selection from a population of variant individuals—individuals having different traits. According to his seminal idea of natural selection, competition within and between species would result in the survival and reproduction of those individuals that were, on the average, fitter than others. As a result, their progeny and—as we now know—their genes would survive. Natural selection is differential reproduction. The extraordinary concept that Darwin put forth was that variation in a population is not just noise but in fact provides the substrate for selection and possible survival.

All of this takes place in evolution over millions of years. But can a selective system work within the lifetime of an individual? We now know that it can: the immune system of vertebrates is a selective system.[2] Your body recognizes shapes of foreign molecules (such as portions of bacteria or viruses or even of simpler organic compounds) through a system of molecules called antibodies. These proteins circulate in your blood and are also present on the surface of the central cells of immunity called lymphocytes.

Immunologists, confronted with the fact that antibodies could bind and even distinguish foreign molecules that never existed before, came up with an instructive theory. It proposed that an antibody, as it was formed, would fold around the shape of the injected foreign molecule (or antigen). The antigen would then be removed, leaving a cavity complementary to its shape. The antibody could then bind to this antigen on future encounters. The idea was beguilingly simple, and it turned out to be wrong.

In fact, it turned out that immune recognition takes place by selection, not by instruction. Within each lymphocyte in your body, the gene for an antibody undergoes variation by mutation and a process called recombination. The result is that the part of the antibody protein that can bind to a foreign antigen on the surface of a given cell is distinctive and unique. Inasmuch as there are as many as one hundred billion lymphocytes, each with one kind of antibody on its surface, a diverse population is formed. When a foreign antigen binds to one or more of the cells via the antibodies that fit its shape, those cells get a signal to divide and produce more of that antibody. The outcome is that subsequent exposures to the immunizing antigen result in speedy binding and neutralization by the much larger number of "specific" antibodies. (I know this system well, having spent a good portion of my research life on this exquisite selectional system and, together with my colleagues, having worked out the chemical structure of antibodies.)

What can we learn from the examples of evolution and immunity? First, we see that there must be a generator of diversity (GOD). Next, there must be a challenge by the environment confronting a species with competition (evolution) or a body with foreign molecules (immunity). Third, there must be differential amplification or reproduction of those variants that are fitter (in evolution) or that fit (as in antigen binding). But note that the mechanisms by which these three principles operate are not the same in the two cases.

We can exploit this conclusion by suggesting that the brain, like the immune system, is a selection system that operates within an individual's lifetime. I proposed this notion in 1977 and elaborated it subsequently under the name Neural Darwinism.[3] The theory has three tenets. The first is that the development of neuronal circuits in the brain leads to enormous microscopic anatomical variation that is a result of a process of continual selection. A major driving force for this developmental selection is the fact that, even in the fetus, neurons that fire together wire together. Two distant neurons will, for example, make synaptic connections if their firing patterns are temporally correlated. Second, an additional and overlapping set of selective events occurs when the repertoire of anatomical circuits that are formed receives signals because of an animal's behavior or experience. This experiential selection occurs through changes in the strength of the synapses that already exist in the brain anatomy. Some synapses are strengthened and some are weakened. It is as if police officers stationed at a particular synapse facilitate signaling from axon to dendrite, while at other synapses, police officers would reduce such signaling. The resultant combinations of signal paths that can be followed in the brain are vast in number, as are the neuronal groups that constitute the selected elements.

The net result of developmental and experiential selection is that some neural circuits are favored over others. But since we abandoned the computer with its logic and clock, how do we get *coherent* behavior out of the system? And what biases the system to yield adaptive responses? The answer to the first question lies in the third tenet of the theory, which proposes a process called reentry.[4] Reentry is the continual signaling from one brain region (or map) to another and back again across massively parallel fibers (axons) that are known to be omnipresent in higher brains. Reentrant signal paths constantly change with the speed of thought (figure 2).

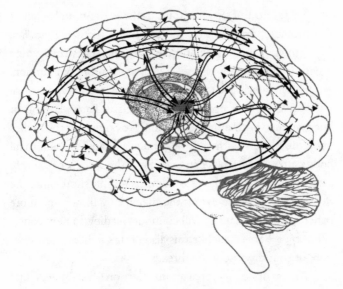

Figure 2.
Reentry is illustrated here by interconnections within the
thalamocortical system. The anatomical arrangements include a
dense meshwork of reciprocal connectivity between the cortex and
the thalamus as well as among different cortical areas. The diagram
cannot even begin to indicate the numbers and density of the reciprocal connections seen in the real brain. These reciprocal
connections, as they carry action potentials and modify
synaptic strengths, integrate and synchronize the
different activities of various specific brain areas.

A net effect of this reentrant traffic is the time-locked or synchronized firing of neuronal groups in particular circuits. This provides the coordination in time and space that would otherwise have to be assured by some form of computation. To help imagine how reentry works, consider a hypothetical string quartet made up of willful musicians. Each plays his or her own tune with different rhythm. Now connect the bodies of all the players with very fine threads (many of them to all body parts). As each player moves, he or she will unconsciously send waves of movement to the others. In a short time, the rhythm and to some extent the melodies will become more coherent. The dynamics will continue, leading to new coherent output. Something like this also occurs in jazz improvisation, of course without the threads!

The theory of neuronal group selection (TNGS) or Neural Darwinism needs one more provision to answer the question about adaptive responses: for successful adaptation, some bias must regulate the outcome of developmental and experiential selection coordinated by reentry. It turns out that, in each species, this bias is inherited in the form of value systems present in the brain as a result of natural selection. Each of these value systems releases a type of neurotransmitter or neuromodulator under particular circumstances. One example is the so-called locus coeruleus, a small collection of neurons on each side of the brain stem. These neurons send their axons into the brain and spinal cord (distributed somewhat like a hairnet for the brain). On receipt of salient startle signals, say a loud noise, these neurons release a neurotransmitter called noradrenaline into the surrounding space, as if from a leaky garden hose. The result can lower the threshold of synaptic responses of multiple neurons, leading to more firing as well as to changes in the synaptic strengths among these neurons.

Similarly, there is a value system that releases the neurotransmitter dopamine. This system is found in the basal ganglia and the brain stem (see figure 1).[5] The release of dopamine acts as a reward system, facilitating learning. Other systems release different neurotransmitters: those releasing serotonin can govern mood, and those releasing acetylcholine can alter thresholds in waking and sleeping. The combination of value system activity, along with the selectional synaptic changes in specific networks of neuronal groups, governs behavior. Selection within these networks determines the categories of an individual animal's behavior; value systems provide the biases and rewards.

We now see that brains have a generator of diversity (GOD), encounter signals from an unknown world through their repertoires of neuronal groups, and facilitate differential amplification of the connections of those groups of neurons that are adaptive. We conclude that our brains are clear-cut examples of selectional systems. Notice that, given the tenets of Neural Darwinism, each brain is necessarily unique in its anatomical structure and its dynamics. Even the brains of twins will differ.

I won't discuss the evidence supporting Neural Darwinism.[6] Instead I will simply state that many experiments have revealed the variance in developmental selection, the importance of synaptic strength changes in learning and memory, and the contribution of reentry to coordination of the activity of brain regions through synchronization of their circuits.

In the view of Neural Darwinism, multiple functionally segregated brain areas such as the cortical regions devoted to vision are bound in their responses by reentry. Cortical area V_1 is dedicated to the orientation of a stimulus, area V_4 to its color, and area V_5 to its motion. These and a score of other areas have no supervisor. Instead, they are reentrantly interconnected by reciprocal fibers (see figure 2). Combinations of responses among these areas give rise to a unified percept, for example, of a tilted, red, cylindrical moving object. This percept arises from the activity of synchronously firing circuits that bind the responses of the various segregated regions together.

According to the theory, memory of such an event is a dynamic system property in which synaptic strengthening and weakening enhances the reengagement of some of the original circuits. But now there is no signal from the original object. Instead there is stimulation, within the brain of a subject, of reentrant circuits to yield an image or thought of the object upon memory recall. In this case, the image is brought up by means of the brain speaking to itself. Memory, which is recategorization influenced by value systems, trades off ultimate precision for associative power.

One final concept is necessary to account for such associative recall: brain circuits under selection must be degenerate. Degeneracy refers to situations in which different structures can yield the same output or consequence.[7] A good example is the genetic code; each triplet of bases in DNA specifies a particular one of the twenty amino acids that go to make up proteins. Since there are four chemically different bases, there are sixty-four possible triplets. However, since there are only twenty different amino acids, the code must be degenerate. Any of the four bases (G, C, A, or T) may occupy the third position in each triplet, in many cases without changing the amino acid specified. There are on average about three (sixty-four divided by twenty) ways to code any one amino acid. So if a string of three hundred bases specifies a sequence of one hundred different amino acids making up a protein, any of roughly three to the power of one hundred different base sequences can specify the same protein sequence. The code is degenerate.

Degeneracy is seen at many levels of biological organization, ranging from properties of cells up to those of language. It is an essential property of selectional systems, which would be likely to fail without it. So we may expect that, in perception and memory, many different circuits of neuronal groups could and do give a similar output. If one circuit fails to function, another is likely to work. The significance of this observation goes beyond the "fail-safe" properties of degenerate circuits. Degeneracy in brain circuits leads almost inevitably to association, a key property required for memory and learning. This associative property occurs because of the overlap of different degenerate circuits leading to a similar output. If input signals change, the existence of that overlap can also result in association with different circuits having different outputs.

A selectional theory such as Neural Darwinism necessarily posits enormously diverse repertoires of neuronal groups. It explains how combinations of such groups can be bound into integrated wholes depending on diverse inputs from the body, the world, and the brain itself. As we shall see, these are just the properties needed to account for the enormously rich yet unitary properties of the conscious state.

four

From Brain Activity to Consciousness

Consciousness reigns but doesn't govern.

— PAUL VALÉRY

WE MAY NOW PROVIDE A PLAUSIBLE account of how consciousness arose in evolution and appeared during individual development in certain species. The evidence suggests that consciousness is entailed by reentrant activity among cortical areas and the thalamus and by the cortex interacting with itself and with subcortical structures. The theory proposes that primary consciousness appeared at a time in evolution when the thalamocortical system was greatly enlarged, accompanied by an increase in the number of specific thalamic nuclei and by enlargement of the cerebral cortex.[1] A starting point for these evolutionary events was probably the transitions from reptiles to birds and separately to mammals about a quarter of a billion years ago.

An animal that evolved with a degenerate reentrant set of circuits linking many cortical regions together could make enormous numbers of discriminations and distinctions. For example, it could link numerous sensory signals together, make many perceptual categorizations, and connect them in various combinations to memory. In this view, primary consciousness emerges from reentrant activity linking perceptual categorization to value-category memory. The pattern of integrative activity in this thalamocortical reentrant neuronal network, called the dynamic core, would create a scene in the remembered present of primary consciousness, a scene with which the animal could lay plans. Clearly, inasmuch as it can make these plans, such a conscious animal would have an adaptive advantage over animals lacking similarly enhanced discriminatory capabilities. The memory system in such a conscious animal is influenced by value systems and by selected synaptic changes brought about by previous categorical experience. This memory system is likely to be mediated by more anterior cortical regions such as frontal and parietal cortex (see figure 1, *top*), whereas ongoing perception is likely to be enabled by more posterior cortical regions.

Consciousness is a process that consists of an enormous variety of so-called qualia: the discriminations entailed by the widely distributed and highly dynamic activity of the thalamocortical core. In such activity, the brain speaks largely to itself. I must stress that it is the *interaction* of the various systems in the core that is critical. We must therefore be careful to avoid assigning consciousness to a specific region.

The understanding that it is the selectional reentrant activity of groups of neurons in the core that yields phenomenal consciousness makes it unnecessary to invoke dualism. Even though consciousness is a process without causal powers itself, it is faithfully entailed by the complex activities and causal powers of the neuronal groups that make up the reentrant core. Moreover, from very early developmental times, signals from the body to the brain and from the brain to itself lay the grounds for the emergence of a self. That self, like consciousness, is also a process. It relies on conscious experience for reference to its own memories, and that conscious experience enhances communication with other individuals of its species.

Of course, in the case of primary consciousness, awareness and conscious planning are limited to the remembered present. An animal with primary consciousness lacks an explicit narrative concept of the past, cannot extensively plan a scenario for a distant future, and has no nameable social self.

For these traits to appear, another evolutionary event had to occur, again involving reentrant connections. At some time in high primate evolution, a new set of reciprocal pathways was developed, making reentrant connections between conceptual maps of the brain and those areas capable of symbolic or semantic reference. We know that symbolic tokens can be mastered by training chimpanzees, and thus chimpanzees with some semantic capabilities may possess the glimmerings of higher-order consciousness. But higher-order consciousness had to await its full flowering during human evolution when true language appeared. At that point, consciousness of consciousness became possible. Reference could be made to a lexicon, the tokens of which could be linked through syntax. Rich concepts of the past, of the future, and of a social self emerged. Consciousness was no longer limited to the remembered present. Consciousness of consciousness became possible.

According to Neural Darwinism, reentry in the enormously complex dynamic core distributed to the thalamus and across the cortex was the key integrative event that led to the emergence of conscious experience. That experience reflected the enormous powers of discrimination made possible by different complex core states. These states necessarily involve integration of multiple aspects of a unitary scene. New core states and unitary scenes differentiate in time and develop as a result of myriad signals from the brain itself, the body, and the world.

In considering this picture, we must not forget that much of behavior is determined by nonconscious interactions among subcortical parts of the brain and the cerebral cortex. Many of these nonconscious responses underlying habits and learned behavior had to be established previously, however, by conscious distinctions mediated by the core. Interactions among core systems, nonconscious memory systems, and signals from value systems operate together to account for the richness of human behavior.

A brief summary of the discussion so far may serve to focus the view of consciousness I am taking here. Conscious states are unitary but change serially over time. They have wide-ranging contents and access. Their range is modulated by attention and, although in large measure they show intentionality—they are about objects or events—they do not exhaust the domains to which they refer. Above all, they entail subjective feelings or qualia. My thesis is that the evolution of a reentrant thalamo-cortical system capable of giving rise to the dynamic core allowed the integration of vastly increased complexes of sensorimotor inputs. Animals having such a core were therefore capable of refined discriminations. Qualia are just those discriminations, each entailed by a different core state. In brief, conscious states reflect the integration of neural states in the core.[2]

With this picture in mind, we can clear up a number of logical errors or semantic inconsistencies that have plagued studies of consciousness. One such error is the failure to distinguish physical causation from logical entailment.[3] Proposals that the action of the thalamocortical core *causes* consciousness confront a difficulty. Since causes precede effects, these proposals imply the existence of a temporal lag between incommensurable processes. Instead, neural action in the core *entails* consciousness, just as the spectrum of the hemoglobin in your blood is entailed by the quantum mechanical structure of that molecule.

Another set of errors is related to this issue of causality. Philosophers have put forth the notion of a zombie, a creature lacking all consciousness but able to behave in all respects as if it were conscious. Possibly, this misguided idea originated from observing the behavior of humans with so-called psychomotor seizures, who can carry out complex acts without conscious awareness. But such act sequences were first learned *consciously.* In the midst of such a seizure, a person cannot learn a *new* task, a task that, like most, requires consciousness for its acquisition. The failure of logic involved in positing a zombie may be revealed by positing a "hemoglobin zombie." Imagine a zombie with all your bodily structures and functions but with blood cells carrying hemoglobin that is white instead of red and yet, even though white, binds oxygen identically to yours. Not likely!

An additional set of confusions may arise because of the reification as things, of properties and processes. Consciousness is not a thing, it is a process. The question "Do qualia exist?" carries the burden of a similar error. A further error is contained in the assertion that sensory categories such as color and various other perceptions exist in the world, independent of mind and language.

Speaking of mind, one often hears of the assumption that a dynamic, self-organizing system like the brain must possess some constant component, or essence, or contrariwise, some sharp temporal or spatial boundaries in order to answer the question "Am I the same self as I was before?" Continuity does not imply essence, nor is it necessary that a system be constant to maintain a resemblance to previous states.

It has often been maintained that, if a structure or property exists, it must "have" a function. This is not necessarily true. What, for example, is the "function" of dreams? Sigmund Freud made an extensive argument for the function of wish fulfillment.[4] But it may simply be that dreams are entailed as a particular state of consciousness when there is blockade of both input and output in the thalamocortical system during rapid eye movement (REM) sleep, a sleep period rich in dreams.

Perhaps the most flagrant breach of logic is the claim that, to explain a phenomenon, one must necessarily replicate it. If you insist on this absolutely, you can never explain consciousness, history, flying, or hurricanes. There are certain processes, however, that must be experienced subjectively before they can be explained. Consciousness is one of them; it is necessarily private because it is entailed by the reentrant core activity in the individual brain.

With some of the underbrush cleared away, we can now explore certain features of brain activity that affect how we obtain knowledge with the goal of providing grounds for a brain-based epistemology. To do so will require us briefly to explore various approaches to how knowledge is obtained and assessed.

five
Epistemology and Its Discontents

*Doubt everything or believe everything: these are two
equally convenient strategies. With either we dispense
with the need for reflection.*

— HENRI POINCARÉ

EPISTEMOLOGY IS THE BRANCH of philosophy concerned with the nature, scope, and origins of knowledge. In brief, it is the theory of knowledge. As such, it has occupied a central role in the development of philosophical thought. But any foray into its branches will reveal a very broad spread of opinion about the validity of ideas in the field and even severe doubts as to the usefulness of their pursuit within philosophy itself. A quick look into epistemological compendia will reveal how the term itself has been subject to modifiers—"feminist epistemology," "virtue epistemology," "traditional epistemology," "naturalized epistemology," and even the "death of epistemology."[1]

I have no intention of plumbing the depths of this contentious field. But inasmuch as my aim in this book is to connect brain science with human knowledge, I have to do some more brush clearing. It will turn out that, looked at from a scientific standpoint, the theory of knowledge is far from complete. To put the enterprise in perspective, I shall briefly review some central issues, ending with issues related to brain-based epistemology.

Traditional epistemology is concerned with knowledge as justified true belief. Much of the philosophical debate about this concern turns around the meaning of the terms "knowledge," "true," and "belief." In that respect, it can be considered to be a language game as described by Ludwig Wittgenstein, who had his doubts about the whole enterprise.[2] The central concern of traditional epistemologists goes back at least as far as Plato's essentialist ideas.[3] In modern times, it can be traced to René Descartes's notions of the solitary thinker searching for beliefs that cannot be doubted. His central notion "cogito ergo sum" was of course the origin of dualism, a metaphysical position that most modern scientists reject.[4] He was centrally concerned with the removal of doubt, with establishing a secure foundation of knowledge. The foundationalist position stemming from the Cartesian view is to some extent the starting point for concerns of traditional epistemologists with the formal nature of mental operations. Their position is also concerned with the normative aspects of the subject as so construed: How can we justify or validate true belief?

The entire position of traditional epistemology turns around arguments related to the thinking subject and the separate world that subject must confront. It is no surprise, then, that debates have occurred among rationalists who emphasize innate mental operations, empiricists who claim that knowledge is achieved mainly from sense data upon interaction with the world, and Kantians who approach the issue by connecting a priori with a posteriori ideas.

Claiming that none of these views reflects the interaction of humans with the world in which they act, a number of thinkers have rejected the entire enterprise. These views may be seen in the essays of Richard Rorty and Charles Taylor, both of whom can reasonably be classified as belonging to the "death of epistemology" school.[5] A central claim of these proponents is that we are not detached observers of the world, operating through "representations" in our mind. Instead, we are agents embedded in the world, gaining our knowledge through action in the world. Moreover, our brains are embodied and that embodiment is essential to any construal of how those brains work to obtain knowledge.

I will not elaborate further on traditional epistemology or the thoughts of its detractors. Their efforts are armchair operations, whatever the correctness of some of their views. Instead, it may be more fruitful to consider efforts to relate epistemology to the scientific enterprise, to naturalize it.

As I have already discussed, a key example is provided by Quine's proposal to naturalize epistemology. His original proposal was that epistemology is concerned with the foundations of science. Aware of the failure of foundationalism, he proposed that we consider the *psychological* processes that lead to beliefs about the world. Quine suggests that the relation to the physical domain should include physics as well as the sensory receptors of the human subject. He suggests that this allows us to maintain the "extensional purity" of physics. The subject receives "controlled input—certain patterns of radiation . . . and delivers as output a description of the three-dimensional external world and its history. The relation between the meager input and the torrential output is a relation we are prompted to study . . . in order to see how evidence relates to theory."[6]

Effectively, this proposal casts epistemological issues in causal terms. But in limiting the field of interest to the world, the skin, and various sensory sheets, the proposal avoids explicit consideration of transformations that occur within the human subject: consciousness, intentionality, memory—all subjects that I have discussed in previous chapters. I will consider later how this breach can be repaired. At this point, it is illuminating to take up another effort to "psychologize" epistemology—that of Jean Piaget.

Piaget, at a time before Quine's proposal, addressed himself to what he called "genetic epistemology."[7] By this, he meant the attempt to explain knowledge "and in particular scientific knowledge on the basis of its history, its sociogenesis, and especially the psychological origins of the notions and operations upon which it is based." Unlike Quine, Piaget actually conducted a program of empirical research, one mainly focused on child development. He proposed that there were patterns of physical and mental events (cognitive structures) underlying intelligence, which appeared at particular stages of development. According to Piaget, there are four stages: the sensorimotor (zero–two years), the preoperational (three–seven years), the concrete operational (eight–eleven years), and the stage of formal operations (twelve–fifteen years). In the sensorimotor stage, one sees intelligence in the form of motor actions; in the preoperational stage, one sees the onset of intuition; in the concrete stage, there is logic but only in terms of concrete referents; and in the formal operational stage, one sees the emergence of abstractions.

Piaget stressed several processes of adaptation: assimilation—the interpretation of events in terms of existing cognitive structures—and accommodation—the alteration of cognitive structures to relate to the environment. In a series of ingenious experiments and observations on children over half a century of effort, Piaget sought to show how knowledge is built up. He challenged the position of traditional epistemology as too static, as neglecting the *development* of knowledge. He also criticized the emphasis of traditionalists on validity and justification that rested on the views of an isolated observer. Instead, he claimed that by following the development of the sciences, we can discover the values and norms that regulate those sciences and the knowledge derived from them.

There is little doubt that Piaget was a great pioneer. There are, however, a number of difficulties in his proposals. The first is related to his notion of the strict sequence of stages. Child psychologists have broadly verified many of Piaget's observations but have challenged his somewhat rigid and speculative framework. Indeed, the term "genetic epistemology" is potentially confusing; his studies might more aptly be described as ontogenetic epistemology.

Regardless of terminology, there is a major weakness in the underpinnings of Piaget's biology.[8] For a good part of his long career, he tended (like Freud) to rely on the discredited biogenetic law of Ernst Heinrich Haeckel: ontogeny recapitulates phylogeny. Of more concern is his rejection of neo-Darwinism, which, in some degree, may have been related to his belief in recapitulation. In addition, Piaget insisted on a dubious connection between the sequences in the psychogenesis of the developing human individual and the historical emergence of scientific ideas. It was overreacting to suggest that the historical development of Western science recapitulates the stages of individual human development. Despite these idiosyncratic proposals and his tendency to overreach, Piaget's empirical efforts had an immense influence on our notions of mental development.

Among the more recent attempts to base a naturalized epistemology on psychology, we may note the work of Michael A. Bishop and J. D. Trout.[9] These authors have put forth perhaps the most extensive (indeed, scathing) critique of what they call standard analytic epistemology (what I have called traditional epistemology). In its place, they propose a program to promote excellence in reasoning. Pointing out the fragility of the reasoning process in many sectors, they propose a program of pragmatic, reason-guiding prescriptions to help us to reason better than any alternative program, say of traditional epistemology. To identify successful reasoning strategies, they suggest an epistemological theory called Strategic Reliabilism. This consists of assessing robustly reliable rules, calculating the costs and benefits of a given strategy, and applying judgments about the significance of a problem—the weight of objective reasons for devoting resources to that problem.

This program replaces the notion of justification in traditional epistemology with pragmatic, rule-based tests of practical success in reasoning. As such, it is normative—if a given psychologically based reasoning strategy yields superior outcomes, it should be adopted. Note that this prescriptive suggestion is based on scientific assessment of real-world outcomes. The normative aspect of this set of proposals should not be conflated with an illicit crossing of the is-ought divide. I have eschewed that crossing for ethics and aesthetics for, in those cases, scientifically based criteria and data are largely lacking. The norms of Strategic Reliabilism are supposed to be adopted only if appropriate reasoning results in assessable outcomes.

The psychologically based enterprise I have briefly summarized here is an exercise in empirically based and naturalized epistemology. It shows much promise, but it does not undertake to consider the neural constraints on behavior that we are concerned with here. It is worth pointing out that the two approaches complement each other: we would like to base epistemology on scientifically certifiable outcomes, but we would also like to assess the relevance or irrelevance of the neural underpinnings of those outcomes.

At this point, a brief mention of two other scientifically based fields may be in order. The first, named evolutionary epistemology by Donald Campbell, has two main branches.[10] The first is concerned with how Darwinian selection constrains the means of knowledge acquisition in a given species. This branch of study overlaps the work of anthropologists and efforts to complete Darwin's program. The second, more dubious branch of evolutionary epistemology is concerned with the wholesale application of selectionist thinking to knowledge itself. An often cited early example is Karl Popper's notion that scientific thought is grounded largely on a series of conjectures subject to selection by attempts at refutation. Another example is Richard Dawkins's proposal that ideas are propagated as "memes" that, like genes, can replicate, be inherited, or be selected against.[11]

This last notion has emerged in the field called evolutionary psychology. Evolutionary psychology is a somewhat more cautious application of E. O. Wilson's sociobiology to patterns of behavior.[12] Wilson's original sociobiological proposal to explain behaviors such as altruism as a result of gene action was sharply criticized.[13] Nonetheless, evolutionary psychology continued to put an emphasis on genes as the prime units of selection ("selfish genes" in Dawkins's designation) in order to explain behavior, particularly social behavior.

In both evolutionary epistemology and evolutionary psychology, a kind of panselectionism is used to explain knowledge and behavior. In sporadic cases the concepts of these fields have some value. But both are attempts to reduce behavior and knowledge to one overarching paradigm. This runs the risk of generating just-so stories that are difficult to verify: logic and propositional analyses are not simply products of evolution and certainly not of selection in present-day individuals or populations. Moreover, genes are not, in general, the units of selection during evolution—individuals are. If Piaget overstepped his bounds in rejecting Darwinian evolution, practitioners in these two fields have overapplied selectionism in their efforts to explain complex entities. Nonetheless, at the level of brain function, selectionism remains a valued approach, one that provides the ground for constructing a brain-based epistemology.

six
A Brain-Based Approach

*Your theory is crazy but it is not
crazy enough to be true.*

— NIELS BOHR

WE CAN NOW ASK THIS question: Can we develop a brain-based epistemology that can begin to deal with the issues raised in the previous chapter? As I have stressed, such an effort must go beyond Quine and deal with the physical bases of mentation and consciousness. It must also be broadly consistent with the developmental emergences first studied by Piaget. As I have attempted to show in previous chapters, the extended theory of neuronal group selection and the analysis of the neural bases of conscious experience are attempts to fulfill these requirements. Here I want to consider further the relation of these matters to the strengths and limitations of a brain-based epistemology.

Quine and Piaget both regarded epistemology as a branch of psychology, not neuroscience. We may ask whether the deficiencies of their proposals may be repaired by our understanding of how the brain works. What can a brain-based epistemology contribute to the picture of how we acquire knowledge? Such an epistemology must be based on evolution—that is, natural selection. This is fundamental, but it must not be mistaken for the assumptions underlying the field of evolutionary epistemology that I mentioned above. It simply refers to the fact that all of the brain mechanisms we have discussed arose during the evolution of *Homo sapiens*. This may seem trivially obvious, but it has some profound implications. One is that the brain, as a fundamental structure for elaboration of knowledge, was not *designed* for knowledge. Evolution is powerful and opportunistic, but it is neither intelligent nor instructionistic.

We can agree with the critics of traditional epistemology that there is no detached Cartesian observer. Instead, the evolutionary assumption just made *necessarily* requires that the brain and body are embedded in the environment (or econiche). And, as we shall see, once language emerged in human evolution, our knowledge and its development, as well as our evolutionary path, depended on culture. As Peter J. Richerson and Robert Boyd point out, culture is not equatable directly to the environment or econiche. I shall return later to this issue.[1]

I have already suggested that the human brain is a selectional system, not an instructional one. The brain of *Homo sapiens* evolved to its present size very rapidly, increasing by a factor of three in size from *Australopithicus* some three and a half million years ago to present-day humans. The largest contribution to this increase was made by the growth of the prefrontal cortex, the part of the brain that is essential for judgment and planning. The evolution of reentrant connectivity provided vertebrate, mammalian, and finally human brains with the most important organizing principle for the acquisition of knowledge. Inasmuch as a large portion of brain development is stochastic and epigenetic—that is, is strongly influenced by the fact that neurons that fire together wire together—no two brains, even those of twins, are identical. Thus, in analyzing the structure and function of the human brain, detailed history must be taken into account, first during evolution and then during individual brain development.

The epigenetic and historical changes in the formation of brain maps are strongly affected by the signals from the body and the environment. This is true during fetal development as well as in development after birth. For example, the proprioceptive system of a late human fetus will distinguish self-generated movements from those imposed from outside. After birth and during infant growth, enormous selectional changes will occur in the synaptic populations of the central nervous system. These changes reach high points in the critical periods of development. In such periods, we find, for example, changes in the first few years when signals from the two eyes lead to ocular dominance columns, structures responding to input from either the left or right eye, thus allowing stereoscopic vision. Later on, as adolescence approaches, the ability to learn multiple languages, previously facile, appears to diminish. Both of these changes are accompanied by vast alterations in the distributions and strengths of synaptic connections. Indeed, even after the major outlines of adult neuroanatomy are established, the borders of cortical maps can change dynamically, depending on the input from the body and the environment. The somatosensory maps of the cerebral cortex mediating touch are a classic example. Increased input from the receptors of specific fingers will not only result in expansion of the cortical somatosensory areas responding to input from these fingers but also cause boundaries to shift for the whole hand.[2] Thus, for example, the cortical maps responding to the left-hand inputs of violinists are greatly expanded.

This dynamic yet historical view of brain development is in accord with the theory of neuronal group selection. I have reviewed it briefly here to emphasize the plastic nature of brain development, which may be considered never to stop until we die. Not only is the fine structure of each brain unique, but the principles of Neural Darwinism lead directly to the notion of degeneracy: different brain structures can carry out the same function or lead to the same output.

What do these observations imply for epistemological concerns? First, the historical, epigenetic, and degenerate features of the vastly complex human brain depend on bodily and environmental inputs and, above all, on action. In the original formulation of the theory of neuronal group selection, it was pointed out that perceptual categorization itself depended on so-called global mappings. These are complex structures composed of both sensory and motor inputs and outputs. The theory states that sensory and motor systems are both necessary to develop perceptual categories.

Second, the notion of reentry as essential for brain development and function puts the emphasis not just on action but also on interaction of brain areas. In a selectional brain, memory, imaging, and thought itself all depend on the brain "speaking to itself" by reentry.

A Brain-Based Approach

Third, by applying the principles of Neural Darwinism, we may resolve the mystification surrounding consciousness and thus extend the ground of naturalized epistemology. Consciousness appeared in vertebrate evolution when reentrant connections in the thalamocortical system arose to link anterior memory systems dealing with value to the more posterior cortical systems devoted to perception. The result was an enormous increase in discriminatory power resulting from myriad integrations among the reentrant circuits comprising this dynamic core. Qualia, entailed by these neuronal interactions, are those various discriminations. Animals equipped with such dynamic discriminatory brain structures had an obvious advantage, particularly for the planning of adaptive responses in food gathering, mating, and defense.

Fourth, being selectional systems, brains operate prima facie not by logic but rather by pattern recognition. This process is *not* precise, as is logic and mathematics. Instead, it trades off specificity and precision, if necessary, to increase its range. It is likely, for example, that early human thought proceeded by metaphor, which, even with the late acquisition of precise means such as logic and mathematical thought, continues to be a major source of imagination and creativity in adult life.[3] The metaphorical capacity of linking disparate entities derives from the associative properties of a reentrant degenerate system. Metaphors have remarkably rich allusive power but, unlike certain other tropes such as simile, can neither be proved nor disproved. They are, nonetheless, a powerful starting point for thoughts that must be refined by other means such as logic. Their properties are certainly consistent with the operation of a pattern-forming selectional brain.

Not only is each such brain unique, but the sensory input from the environment and the motor output of the animal is never identical on separate occasions. This excludes a strict machine model of the brain and mind. It requires that memory be a dynamic, recategorical system property, not a fixed storage of all the variants of a scene, say of a familiar room visited on multiple occasions.

One more essential issue stems from the fact that selectional brains necessarily must operate under constraints imposed by value systems. These are evolutionarily inherited structures in the brain that establish salience, punishment, and reward. As we have already discussed, value systems consist largely of diffuse ascending neural networks that modulate synaptic responses by releasing specific neuromodulators or transmitters in a broadside fashion. One example is the system in the basal ganglia and the brain stem that releases dopamine. Release of dopamine during training is critical to the anticipation of rewarding acts.

Although value systems of this kind are essential, they only constrain actions and perceptual categorization. Value is not category; categorization must be achieved through each individual's behavior. The connection of these notions to issues of emotion and its effect on knowledge is more or less direct. Traditional epistemology rarely if ever concerns itself directly with issues such as emotion, except perhaps obliquely in dealing with normative aspects of justification. In contrast, in brain-based epistemology the mechanism proposed for consciousness by the theory of neuronal group selection is universal: it applies to *all* discriminatory responses whether they involve perception, imagery, memory, feeling and emotion, or even mathematical calculation. In many cases, these processes interact. Brain action cannot be considered, at least at its outset, as a detached process of machinelike calculation in the absence of emotion.

If this picture of principles underlying brain-based epistemology is correct, then early formulations of thought are by nature associatively rich but relatively imprecise. How then do we come to form more precise concepts necessary for scientific pursuits? What about logic and mathematics, both of which involve precision that is essential for enlarging our knowledge and understanding?

Any attempt to answer these questions must confront the issue of language. This is certainly the case for traditional epistemology, which deals largely in propositional or sentential terms. It is also an unavoidable issue in considering the actual development of knowledge and concepts during human history. I have already discussed the onset of higher-order consciousness and its acceleration after the emergence of syntax and a lexicon. The capacity to develop concepts of the past and future and to acquire a social self depends very strongly on the acquisition of language.

We are the only species with a syntax-based language. Many scholars have suggested that language is a biologically evolved trait, and some have even proposed that we possess a specific language-acquisition device that we inherit to allow us to carry out and recognize syntactically correct statements.[4] The theory of neuronal group selection rejects this view. It is surely the case that certain brain regions as well as bodily structures such as the vocal cords and the space above the vocal cords evolved to enhance production and recognition of vocal sounds. It is also evident that the portions of the brain known as the basal ganglia were already able to help the cortex regulate and recognize sequences of motor acts. The interaction of basal ganglia with the motor, sensory, and prefrontal areas of the cortex may have led to a generalized capability of detecting sensorimotor sequences, a kind of "basal syntax." Were that the case, a syntax-based true language may have arisen as an invention based on these already evolved capabilities.

Whatever the case, the possession of language, with its enabling effects on cultural transmission, obviously led to an enormous expansion of conceptual power. Although the linguistic expansion and associative powers of metaphor can lead to poetry and imagination, language also makes possible the development of logic. Logic may have its origins in brain events related to the persistence and disappearance of objects, to the development of operant conditioning, and to the learned consequences of motor acts. In sentential terms, it also allows naturalized epistemologists like Quine to define truth by logician Alfred Tarski's notion of disquotation: "Snow is white" is true if and only if snow is white.[5] When logic is developed in its most refined form, it is most general: the true inferences from sentences characteristic of first-order predicate logic are independent of lexical substitution.

The case of mathematics and its relation to language is even more challenging than that of logic. Is language necessary for arithmetic to be developed? The position that it is necessary is known as the strong Whorfian view, after the linguist Benjamin Whorf.[6] There is empirical evidence that preverbal infants and nonhuman primates have the ability to deal precisely with sets containing from one to four members. Moreover, studies of the indigenous Munduruku people in Brazil have revealed that their language lacks words for numbers beyond five. Although these Indians fail in counting and precise arithmetic beyond the number five, they can compare and "add" large collections of objects. These findings appear to exclude the strong Whorfian hypothesis, for this ability to carry out numerical approximation occurs in the absence of linguistic tokens. It has been suggested that this capability may require the activity in humans of neurons in the parietal cortex, specifically those in the intraparietal sulci (shallow fissures separating folds of the parietal cortex). Although this proposal has been challenged, neurons tuned to numerical quantity have been found in the prefrontal and parietal cortices of macaque monkeys.

The results suggest that although language is perhaps not essential for the beginnings of arithmetic, it plays a role in the further emergence of exact counting and arithmetic during child development. Although the Munduruku do not carry out counting, Western children at around age three suddenly appreciate that each counting word refers to a precise quantity.[7] Thus, one might opt for a "weak" Whorfian hypothesis, although this, too, has been challenged and further analysis will be necessary. The great German mathematician Leopold Kronecker once said, "The natural numbers are the only numbers that assuredly exist. They are given to us by the Almighty. Everything else is the work of man."[8] Perhaps, given what we are beginning to understand, the initial gift was limited to the numbers three or four.

Does a picture emerge from our brief review of various approaches to the theory of knowledge? What we have seen so far is that thought precedes language. But once language sets in, an explosion of possible thoughts occurs, and there is a temptation to equate thoughts and beliefs, and even sometimes knowledge, to propositions and propositions alone. Traditional epistemology has yielded to that temptation. In its search to validate true belief, it indulges in a language game. Its

A Brain-Based Approach

goal is lofty and ambitious, but is based on a narrow set of assumptions about the means by which we think and interact in the world. Its models, which are based either on Cartesian foundationalism (implying a dissociated receiver of instruction or information) or alternatively on a Kantian mixture of a priori and a posteriori ideas, do not seem to correspond to the facts. In proceeding without reference to scientific knowledge and experiment, traditional epistemology ignores how knowledge actually develops.

Quine's suggestion to naturalize epistemology addresses this issue. By limiting its extent to surface receptors and physics, however, it excludes the direct consideration of intentionality— the notion that consciousness is generally about objects, even inexistent ones. But intentionality is a critical aspect of how we acquire knowledge. The analysis of consciousness by the extended theory of neuronal group selection proposes expanding the naturalized view to account not only for intentionality but for the relation between physical causation and conscious experience. By considering value systems that have evolved to constrain selectional systems such as the brain, this theory can also relate emotional experience to knowledge.

If a brain-based epistemology seems a sound way to proceed, what does it recognize and how much can it claim? Brain-based epistemology takes account of the heterogeneous sources of knowledge. It recognizes the primacy of natural selection but does not attempt to explain behavior solely in evolutionary terms. Instead, it emphasizes the epigenetic origins of brain structure and dynamics. In this view, brain development depends on action in the world and, as a consequence, each brain is unique. Pattern recognition by the brain precedes logic, and early thought is creative in its pattern making through processes akin to metaphor. These processes are not free of feeling. Indeed, the constraints of value systems essential to the evolution of adaptive behavior make emotional experience a necessary accompaniment to the acquisition of knowledge even after logic and formal analysis supervene at later stages.

This position helps us to understand the origin of perceptual categorization, of concepts, and of thoughts based on interactions between the brain, the body, and the world. It gives a deeper understanding of such processes as imagery and memory, which are essential to the acquisition of knowledge. Finally, by providing a testable model of consciousness, it clarifies the relationship between physics and conscious thought.

In ranging beyond the narrow bounds of traditional epistemology, brain-based epistemology must take account of insights provided by an analysis of illusions, confabulations, and neuropsychological disorders, all of which lead to distortions of knowledge. Advances in neuroscience promise to shed some light on these areas, which are considered in later chapters.

As powerful as brain science is, however, it is subject to limits. Its detailed exploration of how the brain works remains at an early stage. Furthermore, our understanding of how language is enabled by the brain is in its infancy. Language, arguably the most powerful vehicle for the elaboration of knowledge, both enhances and complicates matters. I hazard a surmise: even if we could accurately record and analyze the activity of millions of brain neurons as an individual formulates a sentence, we could not precisely specify the contents of that sentence by reference to neural recording alone. The idea that we might develop a "cerebroscope" capable of doing so is confuted by the complexity, degeneracy, and unique historical causal path of each brain. Nonetheless, through neuroscientific research, we will certainly be able to develop important generalizations about how we acquire knowledge.[9]

There is another limit to the straightforward application of brain-based epistemology, one that relates to normative issues in various cultures. We must avoid the naturalistic fallacy and concede that "ought" does not derive from "is."[10] To nature we add products of our second nature.[11] A fully reductive scientific explanation of that nature and its ethics and aesthetics is not desirable, likely, or forthcoming. Cultural factors play a large role in determining beliefs, desires, and intentions. As Richerson and Boyd have pointed out, human evolution is accompanied by the coevolution of culture, which provides a relatively rapid and powerful means of change affecting the bases of knowledge, feeling, and behavior.[12]

Last, we must recognize that there are different kinds of truth. Science is concerned with verifiable truth. Mathematical truth rests on formal proof and tautology. Quine defines logical truth as a sentence set from which we get only truth when we substitute other sentences for its simple sentences. Historical truths are harder to establish, depending as they do on unique events in complex situations. The tests of such truths are multiple, but as in the case for science, they may be validated (if that is at all possible) by predictions or, as T. H. Huxley pointed out, by a process of "retrospective prophecy"—the analysis of clues similar to that made famous by the inferential feats of Sherlock Holmes.[13]

So how does epistemology stand with all its discontents? To consider it dead appears to me to be excessive. Yet, even if we accept that empirical science offers it a sanctuary, we must admit that our knowledge of knowledge has great gaps. Even if we agree that a scientific basis for epistemology is a most rewarding path, we must currently settle for a mixture of approaches. This is a humbler and more lenient position than the traditional one, but I suggest that it is more fruitful and revisable.

With this as background, we may examine certain serious splits that have occurred in the forms of human knowledge and consider how they may be repaired.

Forms of Knowledge

THE DIVORCE BETWEEN SCIENCE
AND THE HUMANITIES

Two things fill the mind with ever new and increasing
wonder and awe—the starry heavens above me
and the moral law within me.

—IMMANUEL KANT

OUR DISCUSSION OF brain-based epistemology recognized that there are various forms of truth and different criteria for the validation of each form. In addition to the verifiable truth reached through scientific investigation, there is logical and mathematical truth, and there is truth as established in the writing of history and in law courts. There have been many philosophical approaches to deal with the forms of truth ranging from notions of the synthetic a priori to deep analyses of induction, deduction, and mathematical proof.

The position I have taken is that the naturalization of epistemology must account not only for scientific truth but also for the biological origins in human thought and consciousness of the various other forms of truth. At this point, I want to deal with a long-standing split or divorce between science and the humanities (including the so-called human sciences). After tracing some origins of this split, I will propose an approach to resolving it that is consistent with a scientifically based brain theory. But before tracing these origins, I must point out that when I use the word "science" I refer specifically to Western science dating from its origins in the seventeenth century. Of course, scientific pursuits can be traced back to ancient Egypt, ancient Greece, and even dark periods of the Middle Ages.[1] But the split of which I speak arose in the germ with Galileo and Descartes and was explicitly exposed by the philosophical historian Giambattista Vico in the early decades of the eighteenth century.[2]

The historian Isaiah Berlin traces the divorce between the sciences and the humanities to Vico. This relatively unknown figure challenged the views of Descartes and denied that human beings possessed an unalterable essence. Humans make their own history and understand their own doings in a fashion different than that by which they understand external nature. Our knowledge acquired "from inside," our "second nature," differs from that which we develop from observing the outside world. Instead of the Enlightenment view—a single set of principles applied to all knowledge—Vico applied these contrary thoughts and mounted an attack on the total claims made for the new scientific method. As Berlin indicates, a great debate started "of which the end is not in sight."[3]

Vico's thoughts, which became known only much after his death in 1744, challenged the idea that there was only one set of methods for establishing the truth. From the time of Descartes and Francis Bacon through to the present, one can trace a line of thought that, contrary to Vico, holds up the ideal of a unified system of sciences, natural and humane. Instead of listing all the thinkers on this side (the well-known side reflecting the Enlightenment ideal), I shall first emphasize the other stream in the debate, the one that can be traced to Vico. Then I shall contrast this view to the opposing views held by some modern proponents of reductive or unified science.

A key figure is the German thinker and philosopher Wilhelm Dilthey, who regarded the understanding of human beings as an interpretive matter, one within which notions of physical causation have no place.[4] In his work before 1900 (he died in 1911), Dilthey rejected the notion that humans were essentially rational; instead they exercised willing, feeling, and thinking in various combinations. He assigned the disciplines of psychology, philosophy, and history as *Geisteswissenschaften,* or the human sciences. These were to be distinguished from *Naturwissenschaften,* or the natural sciences, which were concerned with the physical world.

In a manner not far removed from Vico's program, he asserted that descriptive psychology stood at the base of the human sciences. Later, he revised this base to include human history itself, particularly in its sociohistorical contexts. Essentially, Dilthey's positions rested on the notion of hermeneutics, the study of interpretation and its conditions by insiders within a historical culture.

Many modern philosophers have pursued one aspect or another of this stream of the debate. There are, of course, other tributaries of this stream. One might include the differences between science and religion and, more recently, the "science wars," in which postmodernists have suggested the extreme position that science itself has no claim to objectivity but is merely another mode of looking at things, not superior in its truth claims to any other mode.

Rather than pursue these parts of the debate in detail, I wish to make one suggestion that must be considered if the various views on either side are to be reconciled. It seems to me that, if Descartes's dualism is maintained, there must *necessarily* be a split—the human sciences on the side of *res cogitans* (thinking things) and the natural sciences on the side of *res extensa* (extended things). This may seem curious, because Descartes thought to ground all knowledge starting from *res cogitans*. Indeed, Vico rejected Descartes's position.[5] Clearly the position on consciousness I have already exposed rejects Cartesian dualism. In one interpretation, we might claim that William James also rejected substance dualism in denying that consciousness was an entity or a thing, suggesting instead that it was a process whose function is knowing.[6]

The strains and dilemmas that have emerged from the split have driven thinkers to extreme positions as well as to penetrating observations. The philosopher Alfred North Whitehead was deeply concerned with the issue and indeed constructed a whole metaphysics—the philosophy of the organism—to get around it.[7] Later, the debate flared up when C. P. Snow wrote that there were two cultures or polar groups: literary intellectuals versus scientists.[8] Without indulging in such extremity, the physicist Erwin Schrödinger pointed out the curious fact that the great theories of physics did not contain or address sensation or perception but simply assumed them.[9]

On the side of science, extreme postures were adopted with as much energy as those expressed by historians and hermeneuticists. For example, schools of psychology derived from John B. Watson and B. F. Skinner put forth the notion of behaviorism, that all mentalistic explanations should be rejected.[10] Some, like Skinner, admitted of mental events but denied mentalistic causes. In the past decade, a view called eliminative materialism has surfaced that actually claims that there are no mental events or processes.[11]

Another philosophical school of thought, logical positivism, proposed in effect that science was the only legitimate form of knowledge. It was "logical" in its dependence on logical and mathematical studies, and it asserted that a priori knowledge of necessary truths could be made consistent with empirical science. Essentially, the claim was that any statements made outside this frame were neither true nor false but meaningless. Unfortunately, there was no way of showing that the assumptions of this school of thought could themselves meet the criteria of meaningfulness. Some of the thinkers emerging from the so-called Vienna Circle, which provided an early impetus to logical positivism, hoped to formulate a completely unified science. The hope of Otto Neurath, for example, was to give sociology a solid scientific status, but he never achieved the dream.[12] Nonetheless, some of his views were cousin to Quine's later notion of naturalized epistemology.

Two other efforts at scientific reductionism have come to the fore in recent times. The most ambitious is one derived from theoretical physics—the hope of constructing a so-called theory of everything (TOE). This is the search for a coherent formal description (essentially mathematical) that would unify all the four forces of nature—electromagnetism, the weak force, the strong force, and gravity.[13] Some claims have been made that, with string theory, we are well on the way to achieving this goal. Unfortunately, there is presently no single verifiable form of such a theory, and in any event, it certainly would not, in Schrödinger's sense, include an explanation of the sensation and perception necessary to understand it.

Another extreme of scientific reductionism, based on biology rather than physics, has been put forth by E. O. Wilson.[14] He claims that once we understand the so-called epigenetic rules by which the brain is formed and works, we will be able to reconcile human behavior, including normative behavior, by applying these rules. Thus, Wilson claims that even ethics and aesthetics will yield to this reductive analysis, which he calls consilience. The term "consilience" was adopted by Wilson from William Whewell, who used it in his tract *The Philosophy of Inductive Sciences* (1840). By this term, Whewall meant the "jumping together" of facts and theory across disciplines to create a common ground of explanation.

Wilson's statement early on is: "Given that human action comprises events of physical causation, why should the social sciences and the humanities be impervious to consilience with the natural sciences? . . . Nothing fundamental separates the course of human history from the course of physical history, whether in the stars or in organic diversity."[15]

The extremism of this position and of those on the other side speaks to the need for moderation and a different form of reconciliation, to which I now turn. In the course of that effort, I will expand on some of the claims that I have skirted briefly in the above account.

eight
Repairing the Rift

Art is the objectification of feeling, and
the subjectification of nature.

— SUSANNE K. LANGER

CAN WE RESOLVE THE ISSUES that have led to extreme reductionist positions on the side of science and to phenomenology, hermeneutics, and proud humanism on the side of the humanities? Can we repair the rift? As I have said before in considering the Cartesian position, one barrier to repairing the rift was the failure to bring consciousness into the worldview—to naturalize it. That has now become possible, and indeed there is mounting evidence from neuroscience that our cognitive capacities arose in the natural order as a result of evolution. Clearly, these capacities did not stem from logic or computation but instead emerged with the appearance of various brain functions including perception, memory, motor control, emotions, and consciousness itself.

The brain itself emerged during evolution from a series of events that involved historical accidents. Since the human brain and its products developed within a historical context, one might say that the tracing of that development must to some extent involve the same methodology as historians use to trace social change or battles. That is, to some extent, true. But the theory of natural selection, because it is buttressed by molecular genetics and paleontology, allows a historical account of brain evolution that is somewhat more coherent than most descriptions of human exchanges in peace or war.

In one of his essays, Isaiah Berlin makes it clear that the concept of scientific history is untenable for a variety of reasons.[1] First, unlike science, history cannot be described in terms of general laws. This does not mean that historians do not rely on general propositions. They rely on multiple facts and on the general texture of experience, often involving common sense. There is, in general, however, an absence of the models that are so frequent in scientific pursuits. Moreover, the logic and hypothetico-deductive method central to science is not often applicable to historical events.[2] Even though some claim may be made for such approaches in the human sciences of sociology and economics, they are not readily applicable to most historical accounts. If science is concerned with similarities and laws, history is equally concerned with unique events and differences that often depend on beliefs, desires, and intentions within a given culture.[3] In considering human affairs, the scholar or interpreter must place himself or herself within the fabric of these propositional attitudes. General history is a mix of disparate elements that can be studied within different disciplines but not in terms of some general law. Moreover, there are normative elements related to morals and aesthetics that are involved within historical descriptions. These issues pose daunting challenges to the historian, who may have to understand and interpret events that occurred in a culture other than his or her own.

Berlin makes the claim that scientific and historical accounts represent different kinds of knowledge. He expresses this difference by contrasting the views of an external observer and an actor, a contrast between coherence and interpretation.

While the gifted historian must be able to describe the doings of people in many dimensions, scientists on their side do not depend for their generalizations upon contact with common human experience. History, in Berlin's view, is not and cannot be a science.

From time to time, individual historians have attempted to overgeneralize historical interpretation. The results can seem ludicrous. Take for example, the efforts of Brooks Adams, Henry Adams's brother. In a book called *The Law of Civilization and Decay,* he attempted to interpret history in terms of the growth and decline of commerce, with less than satisfactory results.[4] In more recent times, one may note the grand efforts of Oswald Spengler and Arnold Toynbee, both of whose syntheses have fallen by the wayside. And even Vico, in his effort to describe cultural stages in history as those of gods, of heroes, and of men, succumbed to overgeneralization.[5]

Not all attempts to describe and construe past events are so grandiose or silly. John Lewis Gaddis, for example, has put forth an excellent account of the methodology employed by historians.[6] He is aware of the contingent, incomplete, and irreversible complexities of historical events. In describing approaches to deal with such events, he justifiably decries the linear, overly simplistic analyses of many social scientists. Effectively, his claim is that the complexity of history cannot be fit by a Newtonian model, and he rejects the notion of reductionism as a means of historical analysis. But then he suggests that what historians do is closer to the procedures of scientists! He bases this claim on the advances made by scientists in complexity theory, chaos theory, fractals, and the like, advances that he feels share the flavor of the historian's methodology.

Unfortunately, the analogy has several flaws. First, although interesting results have been obtained in the analysis of complex systems, scientists are far from having an adequate picture of far-from-equilibrium or irreversible processes. We still lack adequate means for dealing effectively with multicausal processes for which independent variables cannot be discerned. Second, the measurements made in deterministic systems that are chaotic are still *physical* measurements. Although small initial errors in such measurements propagate to yield chaos, they remain quantitative measurements. Historical systems are rarely, if ever, quantifiable in this way. Nonetheless, Gaddis persists in his analogy and respectfully disagrees with Berlin. The methods of historians he so artfully summarizes still remain largely qualitative.

Gaddis makes a defensible claim that there are sciences with a historical flavor. These include cosmology, geology, paleontology, ecology, and anthropology. It is true that scientists in these areas must take account of historical events, and evolutionary theory and natural selection certainly must deal head-on with such events. (One might even consider Darwin a historian!) Moreover, because of their inevitable complexity and limitations on material, fields such as geology and paleontology must deal with incomplete records. Nonetheless, there are powerful scientific theories that *constrain* these fields—astrophysics for cosmology, plate tectonics for geology, natural selection for biology. No such set of constraining theories is available to historians unless one admits a potpourri of weakly based psychological theories—Freudian analysis, socioeconomic models of rational behavior, and the like. Perhaps the closest analogue to Gaddis's suggestion is ecology, where multiple variables recursively interact in complex environments. Indeed, we may conclude that there are bases for calling ecology a soft science. But, even so, ecology can still marshal a set of constraining scientific theories and quantitative methods not available to historians.

If we accept Berlin's analysis rather than Gaddis's, we may ask why the methodologies and aims of science and historical analysis differ. The answer is not hard to find. Historical events are contingent, usually irreversible, and often unique. They involve high-order issues related to cultural idiosyncrasies, linguistic ambiguity, and specific moral or aesthetic constraints. While, as a person, a scientist is necessarily embedded in such a fabric, his or her aim is to supervene over or transcend the accidents of everyday existence and derive a general set of models and laws in whatever subject domain he or she works.

It is of particular interest, however, that these laws themselves do not give rise to science. People pursuing experiments and hypotheses give rise to laws. Science itself, and clearly

Western science, arose within a particular historical context. What factors govern the actual historical emergence of scientific knowledge beginning with men like Francis Bacon and Galileo and going on to the present?

I believe we can help formulate an answer to this question by considering how the brain evolved and how it operates. In the earlier chapters of this book, I mentioned the evidence that the brain and mind arose as a product of natural selection. I concluded that the human brain itself operates as a selectional system with highly variant repertoires of circuits. Subsets of these circuits are selected to match signals from the world of complex events. In a previous chapter, I argued that the brain is not a computer and that the world is not a piece of coded tape. The brain must, in the absence of unambiguous signals, establish regularities of behavior under constraints of inherited value systems and of idiosyncratic perceptual and memorial events. In human beings, such systems and events necessarily involve emotions and biases.

Selectionistic brains themselves show the effects of historical contingency, irreversibility, and the operation of nonlinear processes. They consist of enormously complex and degenerate networks that are uniquely embodied in each individual. Moreover, human brains operate fundamentally in terms of pattern recognition rather than of logic. They are highly constructive in settling on given patterns and at the same time are constantly open to error. This is seen in perceptual illusions as well as in higher-order beliefs. But as shown by the analysis of learning, error correction is usually available in response to appropriate rewards or punishments.

When we consider modes of thought pursued by selectionistic brains, there is a set of relations between pattern recognition and logic that is both contrastive and a reinforcing.[7] A fundamental early mode of thinking that is highly dependent on pattern recognition involves metaphor. Metaphor is a reflection of the range and associativity of enormously complex and degenerate brain networks. It is pertinent that the products of metaphorical thinking can be understood but cannot be proven as can simile or logical propositions. For example, if I say, "I am in the evening of my life," the statement is understandable but not provable.[8]

Language itself reflects the constructive yet inherently ambiguous and indeterminate aspect of this mode of thought. These features are the result of the trade-off between specificity and range in selectionistic systems that necessarily exhibit degeneracy, a subject I shall address in chapter 10. The diverse repertoires of such systems are never perfect matches to the contents of the domains they must recognize. But after selection occurs across a range of variants, refinement can take place with increasing specificity. This is the case in those situations where logic or mathematics can be applied. We conclude that the necessary price of successful pattern recognition in creative thinking is initial degeneracy, ambiguity, and complexity. In scientific situations, however, the subsequent application of observation, logic, and mathematics can yield laws or at least strong regularities. In the case of historical analysis, qualitative judgment and interpretation are usually the most we can achieve.

Although all of our brain functions and cognitive capacities are constrained by physics and can be understood as products of natural selection, not all of these capabilities can be treated successfully by reduction. As a means to repair the rift, the notion of consilience as proposed by E. O. Wilson is untenable.[9] His idea, for example, that normative systems such as ethics and aesthetics can be reduced to explanation by epigenetic rules of the brain is inconsistent both with the nature of these systems and with how the selectionistic brain works. As David Hume pointed out, "ought" does not come from "is." To assume otherwise is to indulge in G. E. Moore's naturalistic fallacy.[10] Looking at the issue from the side of the brain and mind, epigenetic rules cannot satisfactorily cover the rich complexity and individual history of degenerate networks in the brain. Conscious experiences themselves are enormously complex discriminations in a high-order qualia space, as we have pointed out, and each individual's history and set of brain events are unique. Although there are certainly regularities of intentionality and behavior, they are variable, culture- and language-dependent, and enormously rich. Subjectivity is irreducible.

There is a curiously recursive element in this brain-based

account of how knowledge is acquired. To get science, we need history acting on selectionistic brains. Eventually, this allows the reduction of certain physical and chemical events to general laws. The world order or universe follows physical laws. The remainder of individual and historical events must also follow these laws but cannot be fully explained by or be reduced to them.[11] Irreducible or not, we can agree that all these events are scientifically grounded in the natural order. The evolution of brains and conscious minds occurred by natural selection within the framework of physical laws. So the sequence is clear: following the evolution of *Homo sapiens,* the emergence of language and higher-order consciousness allowed the development of empirical science in the service of the verifiable truth. The application of logic in relation to language and observation of the world, and of mathematics as the study of stable mental objects, profoundly enhanced these developments. Nonetheless, these developments occurred within a specific historical matrix that cannot be reduced to them or by them. Moreover, there is no contradiction in the fact that selectionistic brains capable of higher-order consciousness and pattern recognition could create artistic, aesthetic, or ethical systems within particular historical and cultural conditions. We can conclude that there is no logically necessary divorce between science and the humanities, only a tense relation in which science is admitted as a fundamental but not exhaustive or exclusive basis for grounding our knowledge.

This picture, which is a starting point for brain-based epistemology, is considerably looser than the rigorous developments of epistemological issues by generations of philosophers. It does not, however, exclude these rigorous developments. Rather, it relates them to their ultimate origins in natural and neuronal group selection. In contrast to Quine's efforts at naturalization, brain-based epistemology does not stop at the skin or sensory receptors.[12] It includes more than perception. Indeed, it is based on the analysis by Neural Darwinism of conscious states. The neural underpinnings of such states make human knowledge possible.

It is well to recall that even though all our knowledge depends on our conscious states, these states are necessary but not sufficient for learning. Conscious states themselves appear to have many of the characteristics of irreversible, contingent, and fleeting events. They are unitary but change serially in short intervals of time. They have wide-ranging contents and access to stores of memory and knowledge. They are modulated by attention. Above all, they reflect subjective feelings and the experience of qualia. The evolutionary advantage offered by the emergence of the reentrant dynamic core provided its possessor with vast numbers of sensorimotor discriminations. Qualia are just those discriminations entailed by different core states. They can reflect factual verities as well as illusions and are, in all cases, subject to the constraints of neural value systems.

Given this picture, which is consistent with Neural Darwinism, it is no surprise that rich private experience and external historical events should share properties of both contingency and necessity. The underlying historical processes have complexities that rule out simple reduction of all experience to scientific description. The remarkable event remains: thought within such a system led to the scientific revolution and the generality of scientific laws. It is enough to show how both science and history can be comprehended in our picture of the brain. Divorce is not at issue: the processes that give rise to our understanding comprehend both the sciences and the humanities.

nine
Causation, Illusions, and Values

Reality is merely an illusion, albeit a
very persistent one.

.

Science can only ascertain what is, but not what
should be, and outside of its domain value
judgments of all kinds remain necessary.

—ALBERT EINSTEIN

IN OUR ATTEMPTS TO TRACE and complete the Galilean arc, we must not abandon the aim of science. That aim is to possess a value-free, veridical description of nature, one free of illusions. Science, as the physical chemist Jacobus Henricus van't Hoff once said, is imagination in the service of the verifiable truth.[1] If we accept this, we must also accept that there is no necessary or restrictive constraint on how imagination is exercised, provided that observation and experimentation lead to verification.

The admission of consciousness as a proper target of scientific research has a curious consequence. One must find means of analysis that do not deviate from the causal analysis employed in third-person research. At the same time, one must be aware of the fact that consciousness is a first-person affair, displays intentionality, reflects beliefs and desires, and is subject to illusions and abnormalities that are cousins to creative imagination.[2] To see how to handle this situation, we must analyze the causal connections of brain action. We must then reconcile this analysis with the existence of illusions, useful and otherwise.

First, however, let us take stock and review the argument. We have taken the position that, given the selectionistic properties of the human brain, no adequate reduction of human sciences to the brain's so-called epigenetic rules is realizable. The brain operates by selectional matching of its nonlinear variant repertoires with occasionally novel and nonlinear events provided by world and self signals. With the advent of true language and higher-order consciousness, enormous numbers of discriminations can be experienced. The degeneracy and associativity of these discriminations are accompanied by an even more enormous set of combinations and recombinations of states integrated by the dynamic core. These states are not necessarily veridical and, in addition, are often constructive, contingent, and context-dependent.

The mode of thought that results from these operations initially involves pattern recognition and not logic. Because selection in this neural system is constrained by the operation of heritable value systems and perceptually based memory, the system entails intentionality, beliefs, desires, and emotional states. Such a system is as subject to contingent events from within as it is to external contingencies. It can exhibit singular states as well as regularities, and some of these states are experienced as private, irreducible features of subjectivity.

All of these properties are expressed in one degree or another in thinking and in language. Early on in thinking, metaphor can dominate, and even after the application of logic, language is rich with metaphorical expression. Moreover, as Quine pointed out, language itself shows indeterminacy in reference and translation.[3] The ambiguity that is inherent in natural language is not a critical weakness, however. On the contrary, it is the basis of the rich combinational power that we recognize in imaginative constructions. These properties are just what one would expect to result from the operation of a selectional brain.

Scientific insight results when this power is constrained by logic, mathematics, and controlled observations. But not all judgment and thought can be reduced to scientific description. A key example is the area of normative judgment seen in ethics and aesthetics. Hume's argument still holds: "ought" does not derive in any straightforward way from "is."

These limitations on scientific reduction do not mean that conscious activity, language, and issues of meaning derive from some spooky realm of *res cogitans*. By explaining the neural basis of conscious thought, we can in fact reconcile the appearance of all of the rich properties of thought with physics and biology. The result is indeed a form of reconciliation; divorce is not necessary.

To provide a firm base for this reconciliation (and for a brain-based epistemology), we must address a classical question: Are consciousness and "mental events" causal? And, if not, what is the relation between causal brain action and consciousness? The answers to these questions may startle us, inasmuch as they reveal a set of illusions that we must live with.

It is commonplace to talk of mental events or phenomenal experience as if they were causal. But inasmuch as consciousness is a *process* entailed by integration of neural activity in the reentrant dynamic core, it cannot itself be causal. At the macroscopic level the physical world is causally closed: only transactions at the level of matter or energy can be causal. So it is the activity of the thalamocortical core that is causal, not the phenomenal experience it entails. To make the point clear, let us define C′ to be the integrated pattern of neural activity that makes up the dynamic core at a particular time. C′ entails a conscious state we call C and which involves a particular set of discriminations. C′ not only entails C but contributes causally to subsequent C′ states as well as bodily actions. The relationship between C′ and C is faithful and for this reason, in most cases, we can speak of C as if it is causal. Indeed, C states are informative of C′ states. They are our only access to such states, inasmuch as our neurophysiological methods cannot, at present, record the myriad neural contributions that are integrated in a given causal core state.

So we must conclude that our belief that consciousness causes things to happen is one of a number of useful illusions. The usefulness of this particular illusion may be appreciated by considering that we speak to each other in C language. But the underlying neural activity is what drives individual and mental responses. Philosophers have found this set of conclusions to be an expression of epiphenomenalism—that consciousness does nothing. In fact, it serves to inform us of our brain states and is thus central to our understanding. The traditional horror with which epiphenominalism is met by philosophers can be abated once the faithful entailment mechanisms of reentrant core states are understood.[4]

I have called another conscious illusion the Heraclitean il-lusion because it reflects our way of thinking about time and change. Most people sense the passage of time as the move-ment of a point or a scene from the past to the present to the future. But in a strict physical sense, only the present exists. The integration of core states leading to conscious states takes a finite time of two hundred to five hundred milliseconds. This time period is the lower limit of the remembered present. The past and future are, in contrast, concepts available only to higher-order consciousness. Nonetheless, we often think of the flow of time as we do, in terms of the movement of a Heraclitean river. Within this illusion falls the changing sense of duration we all experience under different conditions. Ex-perienced time, unlike clock time, can seem slow or fast de-pending on various conscious states.

These issues may be connected to two others: the useful-ness of conscious discrimination in planning, over times of seconds to minutes, and the temporal relation between the ac-tivity of the core and of brain areas concerned with action and agency. As I have said, conscious states involve integration times of hundreds of milliseconds. But unconscious neural activity leading to action can lead to much faster responses. Many such responses (aside from innate startle responses) re-quire conscious training. After deliberate practice, habitual responses are then mediated nonconsciously and rapidly by subcortical structures interacting with the cortex. Clearly, it is the play between core states, attention, and subcortical responses that provides the basis for complex suites of action and movement.

Connected to the illusion of a causal consciousness and the Heraclitean illusion is the time-honored and much debated issue of free will.[5] In the strict recognition that all physical events have causes, one must conclude that core states as physical events are determined. Nonetheless, when not physically bound or in prison or in the throes of neural disaster, we can honestly claim the ability to do "as we like" or "see fit," illusory or not. It is on this base that we hold persons responsible for acts determined by society's "oughts," and we train our children accordingly in terms of reward and punishment.

These matters are connected to the relation between normative concerns and neural states. We have dismissed the idea that "ought" comes from "is" and have repudiated the naturalistic fallacy. Nevertheless, we have all inherited a set of neural structures, the value systems, that are critical to the functioning of our brains as selectionistic systems. As I pointed out earlier, the function of these systems is to provide species-specific constraints on the manifold of selective events that can occur in an individual. Suckling reflexes, startle responses, and the action of hormonal pathways and of autonomic neural systems affecting our metabolic and physiological states and emotions are essential to our adaptive functioning. However, they must not be confused with the categories that arise after experiential selection under their constraint. Indeed, in humans with higher-order consciousness, the learning of categories can actually modify the set points of value systems. Humans, unlike most animals, have modifiable value. What ensues is not predictable: there are no animal equivalents of saints, who, even under torture can prefer death to renunciation.

So value systems may jump-start the building of oughts in a society but do not directly determine them. Value systems also provide a brain basis for our complex emotional responses. Unlike Antonio R. Damasio in his excellent account of the neurobiology of emotions, I consider that emotions are complex states arising from core interactions with value systems.[6] The C' states that ensue are accompanied not only by feelings and cognitive content but also by the bodily responses that these states cause. The pleasure and displeasure that can arise clearly reflect the activity of modulatory value responses. But just as C' states reflect huge complexity, their interaction with value systems can also lead to enormous complexes of primary and secondary emotions, with and without cognitive concomitants. All such responses are intimately coupled to the cognitive and emotional construction of the process we call the self. In this respect, whatever errors are laid at Freud's doorstep, he must be credited for his exploration of that process and his attempts to understand it.

When we contrast the picture of generativity provided by brain-based epistemology with that of its philosophical precursors, we cannot help but be struck by a startling difference. What we have called traditional epistemology is concerned with justified true belief and the pursuit of truth and truth conditions. By no means should the importance of such concerns be underestimated. But their pursuit is, in the end, a narrow and terminal enterprise largely preoccupied with language, meaning, and logic. It has not been the proper concern of that enterprise to take up motivation (conscious or not) or emotion or pattern recognition per se. Nonetheless, these are all critical to how knowledge is acquired.

Although the biologically based enterprise is considerably less elegant, it can be looked on as prior to and generative of the traditional view. A possible criticism of this conclusion is that it confounds psychology and epistemology. So be it. It is as important to human knowledge to know how knowledge originates in the long course of evolution as it is to understand truth by disquotation. If we concur that the statement "snow is white" is true if and only if snow is white is an elegant way of assuring a certain kind of truth, it is equally important to recognize its biological as well as its social origins. The reasons are compelling: there are various grounds for claiming truth, and these grounds must be placed in relation to their origins. To be restricted to invariance under lexical substitution (a property of first-order predicate logic) is too narrow a fate. Indeed, the development of logic itself must have depended on the cultural consequences of higher-order consciousness. Creative conscious imagination complemented by logic has gone a long way in the development of scientific truth. It is therefore illuminating to ask how creativity in thought and action emerges from the operation of the brain as a selectional system.

ten

Creativity

THE PLAY BETWEEN SPECIFICITY
AND RANGE

The mind of a man is more intuitive than logical, and
comprehends more than it can coordinate.

—LUC DE CLAPIERS, MARQUIS DE VAUVENARGUES

IN DISCUSSING CREATIVITY, I wish to exercise caution and restraint. My purpose is not to discuss aesthetics or the specifics of artistic creation. It is rather to ask how a selectional brain theory provides a useful background for understanding individual and communal acts of creativity. The word "creativity" itself has rich connotations. To be creative is, by dictionary reference, to be original, inventive, expressive, or imaginative. To create is to make, produce, construct, or call into existence. A reference to God the creator is not rare in certain contexts. Less obvious is the implication that a creator has freedom to create, connecting the various aspects of creativity to problems of free will.

As I said, I want to stay away from these issues, and therefore it may be asked why I turn my attention to the matter of creativity at all. It is because I believe that a case can be made that understanding a brain operating by selectionistic mechanisms to yield a consciousness composed of enormous numbers of discriminations can provide an underlying basis for creative actions. I want to be careful, though, in pursuing this issue. The tenets of Neural Darwinism are not to be considered the proximate or even ultimate explanation of our ability to create in whatever field. But they can cast light on the problem of how conscious and unconscious brain activity can give rise to new ideas, works of art and music, and literary productions. In such works and productions we reveal a second nature. If our scientific description of the world is concerned with nature, our creativity reflects the ability of our brain to give rise to a second nature.

This is the case because of the way in which the complexity of the brain's repertoires can be selectively matched to the complexity of signals from nature itself. I have said that, if the assumptions of Neural Darwinism are correct, then every act of perception is to some degree an act of creation, and every act of memory is to some degree an act of imagination. Remember, in addition, that the mature brain speaks mainly to itself. Dreams, images, fantasies, and a variety of intentional states reflect the massive recombinatorial and integrative power of brain events underlying conscious processes.

Without impinging on the problem of free will, we can certainly see how Neural Darwinism and its extended theory of consciousness provides the basis for such combinatorial activity. First of all, a selectional system must rely on the generation of diversity. The repertoires that result must in general contain very large numbers of variants. A good example to illustrate this point is provided by the immune system.[1] Even though each individual might have the ability to generate different antibodies, if only hundreds or even thousands of variants were produced, the system would fail to recognize the variety of foreign antigens presented by viruses and bacteria. In fact, the number of different antibody variants, one on each separate lymphocyte, exceeds one hundred billion. Beyond some upper limit, however, the cost of producing orders of magnitude more antibodies would yield diminishing returns. Antibody repertoires of adequate size also show degeneracy in their components—a given foreign antigen can be recognized by more than one structurally different antibody. Such a system is not built by information transfer from the objects it ultimately must recognize. Instead, it responds to them by differential amplification of selected variants.

Similar notions are applicable to the brain as a selective system, for neural circuits and dynamics do not *in general* have prescribed information on what the brain will recognize by perceptual categorization. Of course, even in brains, there are evolutionarily determined value systems and reflexes characteristic of a given species. These constrain selection events in response to external and internal signals but do not fully determine them. In a number of its responses the operation of the brain is reminiscent of the statement by E. M. Forster, who reportedly said, "How do I know what I think unless I see what I say."[2]

What has all this to do with creativity? In generating a rich repertoire, only a certain degree of specific recognition can emerge if, in building the system, no information is provided on that which is to be recognized. So if instruction is precluded and yet recognition of a wide variety of states is required, the price paid is a certain loss of specificity. That loss, seen, for example, as ambiguity or indeterminacy in language, is the price that must be paid if the range of signals to be responded to is large. We know, in fact, that the econiche in which animals must survive has an enormous number of signals to which an individual must adapt. For individuals and species to survive, a trade-off must be made between specificity and range.

Similarly, after the range of a brain's or immune system's repertoires is utilized, mechanisms must exist for going beyond the initial selectional steps. The differential amplification of initially selected repertoire elements must be subsequently refined. In the immune system this is achieved by mutating and reselecting already selected cells to produce antibodies with a higher binding energy to the foreign antigen. In the brain, of course, the means of enhancing specificity are quite different.

The brain relies on a number of mechanisms to enhance the specificity of its responses. One involves experiential selection through changes in synaptic strength, constrained by the activity of value systems. The contrast between specificity and range clearly emerges during learning in the change from initial exploratory responses to later conditioned responses. Another source of specificity rests in the mechanisms of attention, which restrict particular patterns of neural response while bypassing others.

The number of possible combinations of corticothalamic patterns of responses is hyperastronomical. The mechanisms mentioned above can be combined with those of short-term working memory (of phone numbers, for example) or of long-term episodic memory of life events in the past to yield outputs that result from repertoires interacting exclusively within the brain.

The important point is that this selective system allows enormous combinatorial freedom for thought and imagery, and even for logic and mathematical calculation. Sequences of thought can be presentational, as in the linkage of visual images, or discursive, as in thinking based on language, where imagery is not necessarily involved. In this account, thought reflects the activity of sensorimotor brain circuits in which the motoric elements are paramount but do not eventuate in action.[3] Although the linkages and sequences during thought involve activity of motor portions of the cortex, the motor cortex itself does not send consequential signals to the motoneurons of the spinal cord or to the muscles.

I have already mentioned my belief that there are two main modes of thought—pattern recognition and logic. I have also suggested that the primary mode, giving enormous range in confronting novelty, is pattern recognition. This is seen primitively in gestalt responses, in the ordering of words, and in various acts of classification.[4] It is enormously powerful, but because of the need for range, it carries with it a loss of specificity. In certain instances, logic can then be employed to eliminate ambiguity. Of course, the use of controlled scientific observation enormously enhances the specificity and generality of these interactions. This movement from range to specificity may be seen as mirroring the generative relationship between brain-based epistemology and traditional epistemology.

We may now return finally to the matter of creativity as a reflection of selectional neural systems. There is enormous freedom for recombination among core states in humans having higher-order consciousness. Creativity in any field must first be permissively allowed for within the huge range of discriminatory qualia. The constraints that are applied through experience and convention prompt various "internal experiments" to emerge, involving order and disorder, tension and relaxation, and the play between the core and nonconscious portions of the brain. Of course, the resulting output is subject to further constraints that come from experiences within a culture. Those experiences determine the choice and response to patterns, altering expectation and prompting abstraction from the flux of experience.

Many of these creative responses depend on the constructive nature of brain action. This can be seen even in the denial of reality that emerges from neuropsychological disorders like anosognosia, which I shall take up next. Unlike the requirement that error must be removed from a computer program, however, the likelihood of error must be tolerated even in the normal individual if the brain is to confront novelty in an adaptive fashion. So it is no surprise that the very origin of Western science depended on the prior existence of certain norms and beliefs that science itself could neither verify nor show to be in error, even after its triumphant emergence as a major source of truth.

eleven
Abnormal States

You must always be puzzled by mental illness. The thing I would dread most, if I became mentally ill, would be your adopting a common sense attitude; that you could take it for granted that I was deluded.

—LUDWIG WITTGENSTEIN

IT IS A COMMONPLACE TO hear that creative genius is akin to or linked to madness. But any consideration of brain diseases that lead to impaired function, delusion, or hallucination will readily limit the truth of that statement. Abnormal states of consciousness, whether the result of drugs, chronic degenerative brain disease, strokes, or the like, unlike creative efforts, do not usually require sophisticated normative standards for us to judge that they are unusual.[1] Victims of neuropsychological syndromes have brain damage that is obviously causal of their symptoms. With psychoses, the issue of causation can be more subtle, and both the etiology and the pathogenesis of disease can be more obscure. Nonetheless, there is no mistaking the advanced delusional or hallucinatory state of a schizophrenic. And although matters can be more elusive in patients with bipolar disorders, their genuine reports of misery and slowed or manic behavior generally deviate from ordinary norms sufficiently to provide no diagnostic obstacles.

With neuroses, however, the question of normative standards comes to the fore. Is unhappiness a neurotic symptom? Are neurotic persons simply suffering from extreme unhappiness? What can we say about these issues?

I propose to consider how our account of brain function, consciousness, and the impact of science on human knowledge may shed light on the issues of abnormal mental states and vice versa. Given the vastness of this subject, I cannot go into great detail. I propose to consider neuropsychological syndromes first because they often illuminate aspects of normal brain function. Then I will briefly consider psychoses, particularly those that are, in one way or another, diseases of consciousness. Last, I will take up the challenging issue of neurosis. It is on this final set of examples that I wish to dwell, for their boundaries with normal behavior are often difficult to discern.

The question I will address at that point is whether one may legitimately entertain or need a theory of neurosis. Obviously, this prompts thought of the predominant exponent of a wide-ranging theory of neurosis and of human personality, Sigmund Freud. Before launching into my thoughts on such matters, I had best set out my response to psychoanalytic theory. First, consider Freud's extraordinary accomplishments.[2] Whatever one thinks of his sometimes exotic metaphors, Freud was the key expositor of the effects of unconscious processes on behavior. Furthermore, even if one rejects his characterization of personality structure, he provided clear descriptions and useful and innovative vocabularies for dealing with mechanisms of so-called ego defense. These are remarkable achievements.

When, however, one considers his accounts of infantile sexuality, of the analysis of dreams, and of repression and memory, matters become harder to evaluate. Freud wove his ideas into a psychological "theory" that could be used to account for neurosis. Unfortunately, the so-called theory consisted mainly of a set of metaphors and was thereby untestable. (As I pointed out earlier, metaphors, unlike similes, can be grasped but cannot be proved or disproved.) The grand sweep of these metaphors was found to be attractive for matters of interpretation in the humanities, and this certainly contributed to the remarkable growth and permeation of Freud's proposals.

Unfortunately, the biological grounds on which Freud based these proposals were ill-founded. An example is his reliance on Lamarckian notions and on the biogenetic law of Haeckel which states that ontogeny recapitulates phylogeny.[3] Moreover, subsequent attempts to test the therapeutic effectiveness of psychoanalysis have been at best inconclusive and at worst unconvincing.

To my mind, Freud's imagination was remarkable and, in certain contexts, even useful, but it did not give rise to scientific insight. His efforts nonetheless provoke the question that I plan eventually to address: Is a scientific theory of neurosis required or even possible? Approaches to answering this question have a bearing on the issues of reductionism that we have been considering all along. Before taking up these approaches, let me turn first to neuropsychology and organic brain disease. This will shed light on those aspects of brain-based epistemology concerned with illusions and the origins of belief.

Neuropsychology as a descriptive discipline goes back to the beginnings of modern neuroscience.[4] Classical examples are Broca's and Wernicke's aphasias. Damage to the motor association cortex extending to the so-called Broca's area leads to impairment of language production. Broca's patients can comprehend spoken language but have failures in language production and show defects in syntax and word order (so-called agrammatism). By contrast, patients with Wernicke's aphasia show major defects in language comprehension. The lesion is in the so-called Wernicke's area near the superior portion of the temporal lobe to which it often extends. Patients with this syndrome have empty speech, in which they fail to express their ideas. They often use the wrong word (paraphasia) or make up neologisms.

Both Broca's and Wernicke's aphasias were described in the late nineteenth century and are classic examples of neuropsychological disorders. Modern research has revealed them to involve more than the cortical areas just described. Subcortical areas are often also implicated. Because of their historical priority, I mention them here as typical examples of functional alterations resulting from the brain damage that is often caused by stroke. Other examples include apraxias (disorders of movements), agnosias (failures to recognize objects despite the ability to see, hear, and so on), alexia (failure in reading skills), agraphia (failure in spelling or writing), dyslexias (difficulties in reading), amnesias (various types of memory loss), and prosopagnosia (failures to recognize faces). To one degree or another, these difficulties (and many others I have not mentioned) can be related more or less well to damage in specific brain areas or to failures in the embryonic development of certain parts of the brain.

In the disorders I have mentioned, brain damage or disordered development clearly leads to certain alterations or losses of function. Each of the disorders I stress here involves a primary compensation, a delusional reaction, a confabulation, or a reconstruction of what undamaged persons would call reality. Sometimes these compensatory reactions are quite subtle and can go undetected. At other times they may appear bizarre. The reason I stress them here is twofold: we know they result from overt brain damage, and the reactions to that damage demonstrate quite clearly the constructive aspect of brain action in the face of severe loss.

A striking example is seen in so-called disconnection syndromes. Most of these involve disruption of the corpus callosum, a commissure (a bundle of nerve tissues) consisting of hundreds of millions of axons interconnecting the two cerebral hemispheres. Rarely, this commissure is not formed or is dissolved in certain genetic diseases. In some cases, patients with certain forms of epilepsy must undergo surgery that involves cutting the corpus callosum. The resulting syndrome has been masterfully studied by Roger Sperry.[5] Although patients with commissurotomy can behave in ordinary circumstances in a fashion that appears normal, Sperry showed that there were distinct behavioral and cognitive differences that could be detected easily by having the patients stare at two screens, one for the right eye and the other for the left. The patient could then respond to commands either verbally or by pointing. As expected, verbal responses to pictures in the right visual field originated in the left hemisphere. But the right hemisphere could respond to an image in the left visual field only by pointing with the left hand. Indeed, in certain tasks of arranging blocks in patterns, the left hand, guided by the right hemisphere, performed somewhat better than the right hand.

In one case, it appeared that a young patient retained the ability to respond to written queries by using the left hand to spell answers with the equivalent of scrabble blocks. Asked about a favorite rock star, the patient using the left brain answered vocally. The left hand, however, spelled out a different name! From similar data, Sperry concluded that there were two consciousnesses at work—that of the everyday person dominated by speech originating in the left hemisphere and that of a more limited type in the right. (This has been disputed as a conclusion, but it has not been disproved.)

What could the origin of two conscious respondents in one body be? The extended theory of Neural Darwinism would propose the hypothesis that there are two dynamic cores, each with capabilities that depend on the reentrant linkages constrained by different target areas in the cortex.

In any event, when the left hand performs in contrary fashion, the vocal patient confabulates or invents a rationalization for any apparent contradiction. This has a bearing on epistemological issues in normal persons. It suggests that the brain of the conscious verbal individual must close a pattern or "make sense" at whatever cost.[6] We will see similar phenomena in other neuropsychological syndromes.

There are other syndromes in which there is a loss of capability on one side of the brain. An example is the syndrome of hemineglect, which sometimes arises when the so-called right parietal cortex (see figure 1) is strongly damaged by a stroke. The patient experiences and reports no vision in the left half of the visual field and behaves accordingly by shaving only on the right side or by reading a clock from twelve to six but not six to twelve.

More extreme strokes, which extend to include areas beyond the parietal on the right, can lead to the curious syndrome of anosognosia. The patient not only has neglect but is completely paralyzed on the left side. But patients with this syndrome deny being paralyzed! They can be intelligent, are capable of normal speech production and comprehension, and are generally not neurotic or psychotic. Nevertheless, they respond to potential contradictions about their performance, asserting that they moved when they didn't by confabulating. And when, some months later, they become aware of their paralysis, their memorial accounts of behavior during the anosognosia period also are confabulations. Once again, it appears that the brain-body interaction combined with past learning results in apparently delusional accounts in which self-consistency is more pertinent than the reports of normal witnesses.

Similarly, in a disease called Anton's syndrome, patients who are physiologically and behaviorally blind assert that they can see. Somewhat related, perhaps, are cases of blindsight in which a patient who is cortically blind in some expanse of the visual field can nonetheless make correct decisions when asked to guess about the identity of figures presented in that blind portion of the field. Certain aspects of previously learned responses can also be seen in patients with prosopagnosia. These patients may not consciously recognize their spouse's face but in a test of pictures containing that face will make positive recognitions in an implicit manner. Sometimes, the conviction of a patient can seem to be truly bizarre. In Capgras's syndrome, for example, some patients show so-called reduplicative paramnesia in which the patient will claim, for example, that his or her mother is in fact not the real mother but is instead an imposter.

I could go on to point out that an epileptic attack in the temporal lobe can result in a patient's conviction that he or she has previously been in a place that is clearly new (*déja vu*) or has had a distinct set of thoughts before (*déja pensée*).

I hope that the examples I have described so far make it clear that the brain-body relationship is critical in the acquisition of knowledge and belief. Most of these syndromes represent diseases of consciousness or attention. Perhaps most important is that the symptoms are not primarily due to early psychic trauma, nor are the patients psychotic. (In some cases, however, psychiatric illness can mimic such symptoms.)

Attempts to account for the confabulation in these syndromes have called on damage to the so-called orbitofrontal cortex.[7] Patients in whom this brain area is damaged or isolated from other connections often show a lack of responsible behavior or an inability to plan actions. Similarly, the so-called mediodorsal nucleus of the thalamus has been associated with spontaneous confabulation. The hypothesis has been advanced that these two brain areas, among others, are necessary for monitoring thoughts that are inappropriate, and when they fail to function, confabulation ensues. What is missing from this hypothesis is an explanation in terms of the interactions of these brain areas with the rest of the brain.

The key conclusion for our purposes is that, even though the detailed pathogenesis of most of these syndromes has not been fully accounted for, no theory beyond a global theory like Neural Darwinism appears necessary for their explanation. Of course, a fascinating challenge remains: when certain brain areas are damaged, to account for how a selectional brain with reentrant thalamocortical interactions responds in the particular way that it does in terms of behavior and beliefs. A rich source of investigation may be to explore whether reentrant interactions are grossly altered in these diseases, so that re-entry among the residual cortical areas may be called on to compensate for loss.

The challenge is, if anything, even greater for another set of abnormal states, the psychoses.[8] Psychoses are diseases in which there is severe impairment of the ability to function in everyday life and to stay in contact with reality. These diseases are especially challenging to our efforts to understand etiology and pathogenesis for, although there is evidence of neural and chemical disorder, there is usually no gross evidence of brain damage. Exceptions exist in cases of toxic psychosis (for example, Korsakoff psychosis in extreme alcoholism), in diseases such as tertiary syphilis, as well as in dementias such as Alzheimer's disease.

But often it is difficult to identify a specific lesion. In the most polymorphous of psychotic diseases, schizophrenia, for example, there is definite evidence for a polygenic mutational origin. But there are also indications of complex environmental contributions, the details of which are hard to disentangle. Patients with schizophrenia show a variety of florid symptoms. These include third-person auditory hallucinations, ideas of reference and influence, and delusions of control by alien forces. These symptoms are often accompanied by emotional blunting and poor interpersonal rapport.

In bipolar disease, a severe mood disorder, there is evidence for a pharmacological imbalance that results in the remarkable slowing of behavioral responses and the conscious depression seen in full-fledged cases. But even given the entire history, it is difficult to attribute the origin of the symptoms to the environment alone. As in the neuropsychological disorders, the suspicion is that no theory of personality is required specifically to account for these abnormal states. Instead, one must identify alterations in brain mechanisms and relate them to scientific hypotheses about brain action. For example, could the hallucinatory and delusional symptoms of schizophrenia be the result of distortions in reentrant timing among higher-order core areas and nonconscious brain regions? If certain temporal delays in linking core responses arose as the result of physiological or microanatomical disturbances, it is conceivable that a patient might confuse his or her own thoughts for exterior voices or for malign references from without. Whatever the case, pathogenetic explanations of the abnormal symptoms of psychosis do not appear to need an elaborate or all-encompassing theory of personality.

Do we need such a theory at all? In asking this question we arrive at the most challenging set of syndromes, grouped rightly or wrongly under the designation "neuroses." It was to the understanding of this domain that Freud's major efforts were directed. Before 1980, neuroses were described as a broad set of diseases, relatively mild as compared to psychoses and without loss of contact with reality. In 1980, in the third edition of the *Diagnostic and Statistical Manual of Mental Disorders*

(*DSM-III*), the American Psychiatric Association eliminated this broad distinction and instead described each disease or syndrome in its own terms.

Neuroses include a number of symptom complexes. Anxiety disorders include anxiety neurosis, phobic neurosis, and obsessive-compulsive neurosis. Hysterical neuroses are categorized under somatoform disorder, comprising conversion hysteria, hypochondriasis, and somatization disorder, and also under dissociative disorder, which covers dissociative hysteria, psychogenic amnesia and fugue states, multiple personality, and depersonalization neurosis.

In phobic neurosis exaggerated fears of events, objects, and bodily functions are seen, none of which is inherently dangerous. Patients show extreme anxiety attached to stimuli from these sources. Specific examples include agoraphobia (fear of public places), acrophobia (fear of heights), and claustrophobia (fear of closed spaces). In obsessive-compulsive disorder, there are recurrent fantasies (obsessions) or actions (compulsions) of which the patient is aware and often tries to resist, but to no avail. There is a bizarre recurrence of notions that the patient cannot avoid, for example, that every third word is "dirty," forcing repetitive covering of the mouth with one hand. The patient knows that this is magical thinking, and thus his or her testing of reality is intact.

In hysterical neuroses, one sees conversion symptoms such as apparent paralysis of certain muscle groups or limbs. In yet others, there are complaints of blindness or deafness. In other cases one may observe dissociation—the loss to conscious awareness of perceptions and memories. This may result in amnesia or in fugue, the loss of identity and recollection of the patient's past life. In somatization disorders, various vague bodily complaints such as headaches, nausea, and the like are present in the absence of actual organic dysfunction. This syndrome can be distinguished from hypochondriasis, in which the patient is convinced that he or she is suffering from a serious disease.

This is no place to detail the rich and extensive proposals that Freud put forth to explain neuroses.[9] According to the ideas underlying psychoanalysis, neurotic symptoms result from inner conflicts. The causes of these conflicts often arose in early childhood from the frustration of infantile sexual drives. The basic postulate of psychoanalysis was incorporated in the notion of a dynamic unconscious mind. In neuroses, a variety of defense mechanisms operate to keep conflicts from conscious recall. Working to achieve conscious recognition of repressed memories became a main aim of psychoanalytic therapy.

Underlying these ideas was an account of the triune structure of human personality consisting of the id, superego, and ego. The id was considered to be a deep domain of the unconscious concerned with the gratification of instinctual drives. The superego was considered to be the result of parental and societal forces acting to censor or repress urges originating in the id. Finally, there was the ego, which mediated between the id and the superego in confronting reality. The ego was in large measure identified with conscious awareness. One of the challenges faced by psychoanalysts was to make the repressed conflicts leading to neurosis available to the conscious mind—"Where id was, there ego shall be."

Freud proposed that dreams were reflections of covert wish fulfillments. Along with the method of free association applied during therapy, dreams were considered to be the "royal road" to the unconscious. In its grand formulation of personality structure, unconscious conflict under repression, and its theory of sexuality, Freud's psychoanalytic edifice had a massive impact on psychology, social science, and literary studies. As I pointed out before, however, the theory failed to meet a number of criteria required for scientific verification.

Freud's constructions, verifiable or not, touch on matters of great import for brain-based epistemology. In the first place, they put great emphasis on symbolization. The mechanisms of ego defense provide a rich dissection of how an individual can avoid confrontation with threats to his or her notion of self. Moreover, the observation that knowledge and experience could be repressed suggested that conscious knowledge and belief were only a small fraction of the cognitive architecture of an individual. This and the felicitous use of suggestive metaphors provided one reason for the influence of Freud's ideas.

Even if we reject certain of Freud's proposals as unscientific and psychoanalytic therapy as barely efficacious, one feature stands out. Psychoanalysis, above all other psychological practices, is concerned centrally with the history of the individual subject, with that individual's personal narrative, beliefs, and styles of thought.

However laudable this effort, it prompts renewal of the question I addressed in considering neuropsychology and psychosis: Beyond a global brain theory, is a theory of abnormal states, in this case, a theory of neurosis, possible or even necessary? Could it not be that what is required is a dissection at the level of brain function of neural mechanisms related to consciousness, attention, automaticity in nonconscious brain areas, and the operation of value systems?

I intend to answer the latter question in the affirmative. Attractive as Freudian metaphors may be, they are formulated in terms that are too removed from the structural mechanisms revealed by studies of brain-body interactions. Although it pays suitable attention to an individual's history, psychoanalysis attempts by means of a general theory to explain that history with all its individuality, irreversibility, and nonlinearity. We have seen how human history resists such general and reductive attempts. Individual history, it is true, differs from human history in the fact that the individual person may, in real time, render a report or record of his experience. But these, we have seen, are subject to a series of necessary illusions. Moreover, as we have seen in neuropsychological syndromes, the brain reacts to defects by filling in and confabulation. In these cases, at least, the problems cannot be laid at the door of psychosexual distortions.

My remarks must not be taken as a rejection of psychotherapy, in the development of which Freud was a pioneer. Given what I have said about contingencies in the history of each individual's development and the ambiguities of language, scientific and causal efforts such as drug therapy need to be accompanied by interpersonal exchange.

What we have seen from our brief review of neuropsychology, psychosis, and neurosis is the extraordinary range of causation and response affecting different levels of brain structure. In neuropsychology, delusions can arise from gross destruction of brain areas. In psychosis, genetic and pharmacological alterations can result in massive distortions of reality testing. And in neuroses, functional connections between thoughts, beliefs, and value system responses can result in disturbed behavior. Here we come across problems relating to the trade-off between specificity and range in a selectional system. Early thought is largely metaphorical, and given its associative powers it can be useful. But if the tension between metaphors in higher-order consciousness and normative values in a culture is given free rein, then it is perhaps not surprising that a rich variety of emotional states and symbolic displacements leading to symptoms can occur. If Neural Darwinism is correct, then even in normal states, every perception is to some degree an act of creation, and every memory is to some degree an act of imagination. The degree is changed in psychiatric illness, and the challenge is to understand how and why.

What about thought and thought processes? Here we may note with approval some of the ideas of Charles Saunders Pierce, the true founder of pragmatism. Pierce pointed out that sensations are immediately present to us as long as they last. He noted that other elements of conscious life, for example, thoughts, are actions having a beginning, middle, and end covering some portion of past or future. This fits our proposal that thought has an essential motoric component reflected in brain action but not in actual movement. Thought, Pierce said, is "a thread of melody running through the succession of our sensations."[10] In contrast, belief is something we are aware of that appeases the irritation of doubt that is the motive for thinking. Belief, he said, is a rule for action or a habit that provides both a resting place for thought and at the same time a new starting point for further thought.

What is needed to understand abnormal states, whether of thought or belief, will be afforded by the validation of a thoroughgoing brain theory such as Neural Darwinism. But this is not enough. If we are to grasp the origins and development of abnormal states, we also must learn more about specific brain mechanisms at all levels, as well as about language and higher-order consciousness. This, and not a reductive theory of personality, has the potential for yielding a fuller grasp of mental pathology and a further understanding of brain-based epistemology.

Brain-Based Devices
TOWARD A CONSCIOUS ARTIFACT

It is clear that there is but one substance in this world,
and that man is its ultimate expression. Compared to
monkeys and the cleverest of animals he is just as
Huygen's planet clock is to a watch of King Julien. If
more wheels and springs are needed to show the motion
of the planets than are required for showing and
repeating hours; and if Vaucanson needed more artistry
in producing a flautist than a duck, his art would have
been even harder put to produce a "talker," and such a
machine, especially in the hands of this new kind of
Prometheus, must no longer be thought of as impossible.

—JULIEN OFFRAY DE LA METTRIE

SO FAR, WE HAVE BEEN concerned with the biological under-pinnings of mental life. On the assumption that we are beginning to understand the brain bases of consciousness, I have explored some of the implications of that understanding for human knowledge and experience. It is frequently but not invariably the case that when science uncovers basic principles or mechanisms, engineering applications based on this knowledge are developed.

Following this line of thought prompts an audacious question that I mentioned in chapter 1: Is it possible to construct a conscious artifact? As I noted in that chapter, the existence of such an artifact would have important consequences for our epistemological concerns. Indeed, of all the consequences of our understanding of the mechanisms of consciousness, a conscious artifact would have the largest effect.

If previous experience is any guide, we can answer our question in the affirmative. We do not know how long it would take to build such a device, but we can consider certain constraints and draw some conclusions about the possible success of such an enterprise. The constraints we must consider are related to the special characteristics of the brain systems underlying consciousness. We must not ignore a number of fundamental properties of the brain. First, we must not forget that the brain is a selectional system. Second, we must pay attention to the observation that the brain is embodied and therefore the brain and the body interact critically with each other.

Moreover, both are embedded in the real world, which obviously has an enormous influence on their dynamics. Third, we know that the reentrant thalamocortical core has enormous complexity involving both integrated and differentiated states. (The unitary scene requires integration in the dynamic core, whereas successive core states are differentiable from each other.) Last, there is a question concerning composition and structure: Would a conscious artifact have to be composed of the same chemical components as are possessed by a conscious human brain?

Before considering how an artifact must meet the major constraints I have just outlined, let me dispose of this last question. The position that the proposed artifact must be made of biochemical components is known as biological chauvinism. The other extreme, which can be termed extreme liberalism, assumes that "hardware," the chemical makeup of the brain, doesn't matter because the brain, being a computer, may run purely as software on a virtual machine. It will come as no surprise that I reject both of these positions. Instead, I claim that it would be sufficient to embed the structures that entail conscious experience, under the constraints I listed above, in whatever material that would adequately meet their functional requirements. The central idea is that it is the overall structure and dynamics, not the material, of an artifact (whether conscious or not) that must resemble those of real brains in order to function. This requirement has already been met by the design and construction at The Neurosciences Institute of a series of brain-based devices (called by their inventors BBDs).[1] Though far from exhibiting conscious behavior, these real-world devices are capable of perceptual categorization, learning, and conditioning without instruction. They are even beginning to display episodic memory, a characteristic of hippocampal function, and as a result, they can autonomously locate themselves and designated targets in a real-world scene.

I will describe these brain-based devices at some length, but first I want to contrast their design with invented machines and with robots and summarize human efforts to use machines and animal species to accomplish various tasks. What will be clear from such a survey is that, in the past, no assumptions about conscious machines were entertained during such efforts. More variably, it may have been the case from time to time that the animals helping humans in various efforts were assumed to be conscious. But this assumption was secondary to the notion that these animals were capable of trained behavior.

From the times of the great pyramids, humans used simple machines and animals. What might be called passive or informational machines were also used, for example in early astronomical observations. Whether passive or active, a machine was a device designed or formed to carry out a specific function or task. Beyond the use of levers and wheeled devices and before the invention of more intricate machines, two animal species, horses and dogs (and more rarely, oxen and elephants for pulling and lifting), were trained to perform tasks of locomotion and herding. After the steam engine was applied to railroads and the Otto cycle engine to automobiles, the horse was only rarely employed in locomotion.

The application of sophisticated machines expanded greatly with the invention of communication devices—the telegraph, the telephone, radio, and television. And, of course, the explosion of applications that emerged from the invention of the digital computer, fueled by solid-state physics and microelectronics, continues to affect our lives.

In some sense, one may conclude that the digital computer, arguably the most interesting invention of the twentieth century, is the quintessential machine. Turing's demonstration that one can envision a universal Turing machine, one that can successfully carry out any sequence of computations based on effective procedures is a spectacular generalization of the idea of a machine.[2]

Why cannot we consider Alan Turing's theories to apply *grosso modo* to the brain? We have already discussed that in brain development, a certain amount of dice tossing occurs; this is incompatible with Turing machine structure. Furthermore, the world confronted by the body and brain is not unambiguous (and so fails to meet the requirement for a sequence of algorithms or effective procedures). Brains must therefore operate ex post facto by selection from repertoires of variants. A brain-based device must therefore also operate by selection as it confronts shifting contexts in the real world.

We may well ask, Why could we not meet the need for behavior in the real world by constructing a robot? A robot can be defined as a "programmable multifunctional device designed to manipulate or transport parts, tools, or specialized implements through variable patterns for the performance of specific tasks."[3] The development of robotics from the late 1940s to the present has been truly impressive, and major industries concerned with manufacturing and control have paid increasing attention to the further development of these devices. The hope, of course, is to construct a fully autonomous robot, one that could successfully navigate an environment to carry out a variety of tasks, including interaction with people, just as horses and herding dogs did in the past. So far this idea has not been realized, although some successful steps have been made. Still, such devices are not based on the brain in any degree and are certainly not based on neuroanatomy driven by selectional principles.

What then are the possibilities of constructing devices based on such principles? Attempts to answer that question have prompted the construction of the brain-based devices that I mentioned above. The motivating idea behind the construction of brain-based devices was a general one related to the problem of determining how the brain works. Clearly the solution of that problem will continue to depend on experiments on living animals of a variety of species behaving in real challenging environments. Just such experiments have led to the enormous expansion of knowledge in modern neuroscience. But there is a limit that is methodological and eventually even ethical: in a single live animal, we cannot simultaneously (or even sequentially) examine *all* brain and bodily events ranging across levels from molecules to behavior. Without this ability, it is difficult and sometimes impossible to grasp the multilevel interactions necessary for understanding the origins of complex neural responses and behavior. If, however, we could build a brain-based device whose workings can be followed in all details, it might be possible to gain insights into multilevel brain events and their interactions with behavior.

This possibility has driven a twelve-year program at The Neurosciences Institute, where scientists and engineers constructed real-world behaving devices that performed autonomously in an environment, guided by simulated brains whose structure and dynamics are based on selectional principles. In this program, a series of brain-based devices has been realized; the successive artifacts have been named after Darwin. To give some notion of their design, I will consider the structure and performance of three of the more recent devices: Darwin VII, Darwin VIII, and Darwin X. (See figure 3 for a photo of Darwin VII.)[4]

The brains of these devices are simulated in a powerful

computer array. Brain responses are transmitted by wireless links to the body or "phenotype" of the behaving device named NOMAD (neurally organized mobile adaptive device). The actual phenotype can be altered to include a variety of sensors present on a wheeled platform capable of exploring an environment having varying characteristics. Most NOMAD platforms contain a charge-coupled device camera for vision, two microphones embedded in structures for hearing, a gripper capable of grabbing blocks with different visual patterns on them, and, in some cases, whiskerlike protuberances capable of distinguishing surfaces with different degrees of roughness The environments in which various Darwins perform consist of rooms ranging from ten feet by twelve feet to rooms as long as twenty feet, with black floors and enclosures that are illuminated by overhead lights. In this type of environment, NOMAD platforms traverse various paths autonomously in response to visual, auditory, and tactile signals.

Figure 3.
The brain-based device Darwin VII has learned on its own to pick up and "taste" striped blocks and avoid blocks with blob patterns. It has a charge-coupled device camera for visual input, microphones on the two "ears" for auditory input, and a row of infrared sensors to avoid collisions. The gripper shown grabbing the striped block can sense conductivity ("taste") and pick up blocks that it encounters in its travels. Its simulated brain is based on vertebrate nervous systems, and it operates by selection rather than instruction.

Brain-Based Devices

The simulated brains consist of definite neuroanatomical areas based, for example, on the mammalian visual cortex, inferotemporal cortex, auditory cortex, and somatosensory cortex (see figure 1). The neuronal units making up these areas are connected synaptically, and the weights of the connections between the various units are generally modifiable according to a rule for modeling changes in synaptic strength. In addition, the simulated anatomical circuitry contains the equivalent of value systems modeled on those seen in real brains. These are essential to constrain the selectional responses of a particular Darwin's brain both to its own signals and to those from the environment.

The neuronal units respond according to a mean firing-rate model; in this model, each unit represents the collective responses of a group of approximately one hundred neurons. These responses drive a series of motor outputs that impel exploratory or reactive movements of the Darwin in question. At this point, let us consider Darwin VII.

Before training, Darwin VII traversed its environment by a series of exploratory moves. When drawn by its visual response, it approached blocks having top portions covered by vertical or horizontal stripes or by blobs. When sufficiently close, the brain-based device would attempt to "taste" a proximal block with its gripper. "Taste" was defined by the experimenter as low conductance (the gripper being able to measure that variable on contact) or as high conductance. Arbitrarily, low conductance led to activity in the value system that resulted in aversive movements, whereas high conductance prompted appetitive movements and grasping of the block by the gripper.

With these constraints and without instruction, exemplars of Darwin VII would initially grasp both "good" and "bad" tasting blocks. Within short order, however, it became conditioned to lift and taste only "good" striped blocks and to avoid "bad" tasting blocks with blobs. If the blocks with blobs emitted a low tone, Darwin VII could be conditioned secondarily to avoid such blocks while approaching, but it would "taste" striped blocks that emitted high tones.

During all of this primary and secondary conditioning, the experimenter could record the totality of responses in Darwin VII's nervous system, which contained twenty-five thousand neuronal groups and half a million synapses. It was found, for example, that an area simulating the inferotemporal cortex would respond with an activity pattern of neuronal groups that was characteristic and relatively invariant in the responses to striped blocks, whatever their position in space. This pattern shifted to a new and different one for responses to blobs. The invariance developed only after complete rounds of actual movement and behavior characteristic of that individual Darwin VII. Another individual starting from the same anatomy would develop its own characteristic and unique "inferotemporal" patterns! The brains of these brain-based devices, like our own selectional brains, developed individual activity patterns that nonetheless tended to result in behavioral responses that were similar among different individuals.

With this background, I will very briefly describe the behavior of two other brain-based devices, Darwin VIII and Darwin X. Darwin VIII resembled Darwin VII, but its nervous system had an additional feature: reentry. The presence of reentrant long-range connections enabled it to distinguish among multiple potentially confusing objects of different colors and shapes. Confronted with green square-shaped objects and red diamond-shaped objects of the same size on one wall and red and green squares on the opposite wall, it was able to turn selectively to a green square on either wall on the basis of having received a positive auditory value signal accompanying its vision of a green square. It was able to make this distinction because reentry enabled it to solve the binding problem—how different brain areas can synchronize and integrate their segregated functions in the absence of an executive area. It correctly associated green color and square shape by the synchronous firing of selected neuronal groups in the inferotemporal area of its simulated brain.

Darwin X incorporated another critical brain region, the hippocampus, as well as many more neuronal groups and synapses. The groups numbered over one hundred thousand with two and a half million synapses among them. The hippocampus in animals is responsible for episodic memory, the

long-term memory of sequences of events. It is also responsible for an animal's ability (for example, a rat's) to locate itself
in space on the bases of cues received from the environment.
Placing a rat in the so-called Morris water maze, for example,
will impel the rat to swim in milky water until it inadvertently
finds a hidden platform upon which it may rest from the apparently unpleasant experience of undirected swimming. Its
hippocampus allows it to remember visual cues from the surrounding walls. On being placed again in any part of the pool,
the rat can swim rather directly to the hidden platform. This
is based in part on the activity of so-called place cells in the
hippocampus, and a series of interactions between the hippocampus and the cerebral cortex.

Darwin X was placed in a similar situation but without
water. Instead there is a black-filled circle of different reflectance than the black floor on which the device moves. The
brain-based device cannot see this circle but can detect it when
it stands over it as the result of an infrared detector sensing the
circle's reflectance and sending a positive value signal to its brain.
Each of the four walls of the enclosure in which Darwin X
moves contains different-colored stripes in distinctive arrangements. After patrolling the enclosure, Darwin X remembers
these patterns. Following the experience of a few episodes of
positive value after coming on the target circle, it will more or
less directly head to the circle no matter where the device is

initially placed in the enclosure. In fact, the device's simulated hippocampus develops characteristic place cells in a fashion similar to those of living animals, for example, rats.

Clearly, the performance of brain-based devices is encouraging, even if their development is still at an early stage. Several aspects of the behavior of these devices must be put in perspective. First of all, they are not robots as defined previously. Their behavior is not preprogrammed according to a fixed sequence of algorithms. It is true that their brains are simulated inside powerful computer clusters, but no target function is prescribed or prespecified and the initial synaptic strengths in the brain are set at random. Instead of having a defined program, the brains of the devices are built to have neuro-anatomical structures and neuronal dynamics modeled on those known to have arisen during animal evolution and development. Furthermore, they are situated in an environment that allows them to make any of a number of movements to sample various signal sequences. Moreover, although their "species-specific" pattern is constrained by "inherited" value, value is not equivalent to category. Instead, the Darwin devices develop perceptual categories on the basis of their experiences in the real world, and they build appropriate memory systems in response.

Notice that the performance of brain-based devices speaks against both extremes of chauvinism and liberalism. These devices are not made of biochemical components. Moreover, they are not just software running a virtual machine. Even though they are not alive, they can carry out conditioning, perceptual discrimination, and episodic memory in a fashion that involves circuit dynamics resembling those observed in the brains of living animals.

This brings us back to the query with which we started: Is it possible to construct a conscious artifact? Is it necessary that to be conscious such an artifact must be living? A living system may be considered as one capable of self-replication that is subject to natural selection. If BBDs (which are presently far from meeting the requirements of conscious systems) are any guide, a tentative answer would be that a conscious artifact would not necessarily have to be alive. Given the presence of a body with sensory and motor systems, what would be necessary is a high degree of complexity in the simulated equivalent of a thalamocortical system interacting with a basal ganglion system. That complexity is presently unrealizable.

Aside from such structural limitations, there is an additional requirement if reasonable criteria indicating conscious behavior are to be satisfied. Such an artifact would have to have a true language, one with syntax as well as semantics. In other words, it would have to have a form of higher-order consciousness. Only with that trait, capable of giving us reports, could we experimentally test brain function with results sufficient to support the conclusion that the artifact was likely to be conscious.

This goal is remote at present. Nonetheless, it is of some interest to consider what an artifact, having the high dimensional discrimination that is characteristic of a conscious brain, would provide in its reports about the patterns and consistencies present in the physical world. Would it report consistencies similar to those of our physics? Or, given its phenotype, would it carve up the world in ways more similar to those seen in neuropsychological disorders?

Whatever the case, if such an entity were devised, it would be fascinating to explore whether it could be incorporated into a hybrid machine, a perception-Turing machine.[5] Such a hybrid would combine the strengths of syntactic machines, such as computers based on human programming, with the semantic abilities of an artifact able to deal with novelty and noncomputable inputs.

I suggest that someday a conscious artifact could probably be built. But it remains a remote goal. Even if that goal is reached, such a device would hardly challenge our uniqueness. Remember that the brain is embodied and that we are embedded in an econiche and culture that could hardly be duplicated or even imitated. The human phenotype with all its complexity is what fuels our particular qualia. The likelihood of matching such a phenotype verges on zero. The precious qualities of our own phenomenal state are safe from preemption or displacement.

Given our present restrictions, we may sidestep an ethical

question: What would human responsibilities be, for example, in removing such an artifact's capacities for consciousness after it has accumulated experience and developed a unique identity. This issue, which certainly does not have to be faced now, relates to the instrumental and moral value of human knowledge itself.

thirteen
Second Nature

THE TRANSFORMATION OF KNOWLEDGE

In my end is my beginning.

—T. S. ELIOT

WE ARE PART OF NATURE AND are embedded in it in ways that Darwin successfully described. But often in our heads we have ways of looking at the world that "feel natural" or "are second nature" even in the face of scientific evidence to the contrary. Do the two natures have to be reconciled? Some, hewing exclusively to the imaginative side or to art, would say, "not necessarily." Others, in the camp of extreme scientific reductionism, would say that ultimately all subjects that are products of the mind should be reducible to explanations based in brain science, to so-called epigenetic rules.

Clearly the path I have pursued here differs from both of these positions. I have searched for a reconciliation by examining how the pinnacle of our mental life, consciousness, arose in the course of evolution. An examination of the biological bases of consciousness reveals it to be based in a selectional system. This provides grounds for understanding the complexity, the irreversibility, and the historical contingency of our phenomenal experience. These properties, which affect how we know, rule out an all-inclusive reduction to scientific description of certain products of our mental life such as art and ethics. But this does not mean that we have to invoke strange physical states, dualism, or panpsychism to explain the origin of conscious qualia. All of our mental life, reducible and irreducible, is based on the structure and dynamics of our brain. Darwin was right in rejecting Wallace's conclusion that the human mind could not have arisen by natural selection.[1] Indeed, as I hope to have shown, a set of evolutionary events produced the neuroanatomical bases for reentry, which led to the development of the enormous number of discriminatory states, or qualia, characteristic of conscious experience.

The methodological limits that are imposed by the complex properties of the reentrant thalamocortical core do not prevent us from studying neural correlates of consciousness. But no amount of experiments on such correlates will in itself provide a basis for understanding how qualia arise. As I hope to have shown, that understanding must be achieved by logical and linguistic analyses within brain-based epistemology.

The adaptive advantages provided by the evolution of the neuronal arrangements in the dynamic core are evident. The activity of such neuronal arrangements allowed an animal to make an enormous number of discriminations among both internal and external states and across a variety of modalities. Qualia, which are those discriminations, differ relative to each other because they originate from the integrative interactions of quite different neural arrangements. Planning of adaptive responses by conscious animals was enhanced by these means. Although we do not know what it is like to be a bat in the same way we know (by experience and homology) what it is like to be human, we may reasonably surmise that in its discriminations, a bat's core state will be as dominated by echolocation as ours tends to be by vision.

A key issue concerns the efficacy of our conscious experience. The problem of relating neuronal action to phenomenal subjective experience is solved by a causal analysis. Qualia are entailed by states of core neurons acting to yield complex integrative states that can shift to yield new states and conscious scenes. Qualia are thus no more *caused* by neural states than is the spectrum of hemoglobin caused by that protein's structure— its so-called Soret spectrum is entailed by its molecular structure. In the case of core systems in the brain, the relationship of entailment is a faithful one even if it is degenerate. That is, everything else being equal, the same core state will not entail radically different qualia.

Qualia are not themselves causal, and to assume otherwise would go against the laws of physics. But there is no need for that to be the case, given the fidelity and causal efficacy of core states that we have called C'. The corresponding qualia that we may call C are informative even if they are not causal. Indeed, at present, because we lack the means of fully detailing the hyperastronomical interactions of core neurons, C provides the only indicator we have of any overall core state, C'. Indeed, our methodological inability to reduce to cellular or molecular terms the mental or conscious events accompanying fields such as ethics and aesthetics that emerge when we speak "C language" to each other should not be construed as arising from the existence of some radically inaccessible domain.

In the view put forward here, even given the irreducibility of certain subjective conscious experiences, we can understand how our second nature arises from scientifically describable foundations. Although it is true that a scientific description of the world hews more closely to the structure of that world than do our daily impressions, our account of how the brain works suggests that scientific hypotheses themselves emerge from ambiguous (and occasionally irreducible) properties that give rise to pattern recognition. The brain structures and dynamics leading to such properties *are* scientifically describable, even if the properties themselves cannot be fully reduced. Similar considerations apply to the cultural exchanges that give rise to art and to ethics, the relationships of which are not entirely subject to rigorous scientific reduction. No limitation of our potential is implied by this view. Creative matching of social experience, developments in art, and expansion of our knowledge in all spheres have no obvious limit.

Globally speaking, scientific observations and theory can provide descriptions of the brain events that result in such activities. The grand sequence—Big Bang, cosmos, galaxies, Earth, origin of life, evolution, mammalian brains, hominid core development, language, Galilean science, relativity and quantum mechanics, modern neuroscience, neural bases of consciousness—may eventually be able to explain the background, if not the details, of individual subjective histories. These histories are recursively embedded in such a sequence and they account for its human products. They derive ultimately from natural selection. This grand view fills in the Galilean arc and helps to complete Darwin's program.

How does this view, derived from brain-based epistemology, differ from that of Quine? Quine, as I mentioned at the beginning of this book, considered consciousness a mystery but correctly took it to be a property of the body. He rejected the Cartesian dream of scientific certainty. Instead he proposed that epistemology be naturalized in terms of stimulation of our sensory receptors by signals from the external world, without denying the senses themselves. In this way, he suggested, quite fruitfully, that philosophical and logical analyses could be linked to science. He limited his consideration to sensory receptors and did not concern himself with mental life, because he felt that introducing intentionality into a scientific theory of the world would destroy "the crystalline purity of extensionality: that is, the substitutivity of identity."[2] In taking this position, he did not, however, exclude the scientific exploration of intentional issues. I suspect that he recognized the limitations in knowledge of his time and that, had he had knowledge of consciousness research as we do, he would have expanded his domain.

Thanks to modern neuroscience, many of these limitations have been removed. Franz Brentano's notion of intentionality, the observation that conscious states are generally about objects or events, can be explained by the extended theory of Neural Darwinism.[3] Consciousness is proposed to arise initially as a result of perceptual categorization in interaction with memory systems. By its nature, such categorization, though unconscious, is necessarily about objects and events.

In my opinion, we do not destroy the crystalline purity of Quine's proposal by going beyond sensory receptors. Rather, we may extend it into domains excluded from traditional epistemology. It is true that the approach I have taken to the subject does not maintain the sharp boundary between psychology and epistemology that is maintained by some. I find this to be an advantage: with this approach, we can usefully consider the origin of logic in language, the contribution of imaginative pattern recognition to mathematics, the historical and ideational origins of scientific empiricism, and all manner of artistic and normative issues. Of course, the boundaries separating all of these domains must be specified. But we will no longer consider the origins of true justified belief to lie only in language, however central language is to higher-order consciousness. Essentially, accepting brain-based epistemology amounts to accepting empirical data from neuroscience as well as psychology to buttress our views of the origin and nature of human knowledge.

That acceptance does not imply that brain-based epistemology is exhaustive or that it excludes scientifically grounded normative functions related to epistemological concerns. The main strength of the brain-based approach is that it provides scientific grounds for a pluralistic view of truth. At the same time, it supplies useful constraints on how we attain knowledge. By including a scientific view of consciousness, it rejects the position that naturalism undermines first-person authority, the illusion of conscious will. Given its position on the origins of language and the effects of culture, the brain-based view has no problem with normative judgments that go beyond the justification posited by traditional epistemology. It is entirely compatible, for example, with proposals that a major goal of epistemology is to set up norms that assure excellence in reasoning.[4]

Truth, though heterogeneous in its origins, is itself normative and thus worth caring about.[5] Establishing truth requires many different means and methodologies. These cannot be traced directly back to evolution or the physiology of the brain.[6] One of the main messages of this book is that, although we must recognize that evolution and neuronal group selection provide the bases and constraints for the acquisition of knowledge, historical, sociocultural, and linguistic factors set up normative criteria for truth.[7] A key point is that these criteria can, by these means, be established in a naturalized fashion.

By analyzing consciousness in scientific terms, Neural Darwinism rejects Cartesian foundationalism and dualism. Physics and biology can live together while recognizing that each has its historical origins in human experience. A word about the relation of that experience to the scientific enterprise may be useful. Following hominid evolution and the coevolution of culture, it became possible to develop experimental and theoretical physics, initially evoked most clearly in the work of Galileo. Scientific observers and experimentalists then developed descriptions that led to the formulation of general laws. It must be emphasized that those descriptions are not the events described. Despite the power of prediction and invention provided by science, it does not duplicate the world.

Furthermore, after Darwin, when it became possible to apply global scientific approaches to brain function and consciousness, a similar limit became apparent: The scientific description is not the experience. Of course, the description of consciousness helps us to understand our experience in a way that physics alone could not do. Nevertheless, it is important to recognize the priority of experience in giving rise to the descriptions that illuminate the bases of that experience itself.

Once higher-order consciousness and language operate recursively to connect thought, emotion, memory, and experience, the number of discriminative combinations grows without bound. We move in corridors of the mind ranging from the certainties of mathematical insight to the fantasies of *A Midsummer Night's Dream*. Often, the parts of our second nature that seem to deviate most from the truth are just those necessary to establish new truths. But of course, they are not sufficient. Various criteria must be applied to establish each kind of truth. The main point is that truth is not a given, it is a value that must be worked for during our personal and interpersonal interactions. The richness of those interactions is no surprise given the associativity and degeneracy of reentrant interactions in the brain.

If, in our scientific descriptions of world events and of consciousness, we do not duplicate either the events or the experience, is that personal experience a form of knowledge? Given the range that is covered by higher-order consciousness and despite its subjectivity, we must admit that qualia are indeed discriminative forms of knowledge. By including myriad possibilities of pattern recognition, metaphor, and complexity, such knowledge goes beyond the formalities of justified true belief.

If we adhere to this language game, we must qualify the connection between knowledge and truth. In this view knowledge and truth are not the same. Taking this position would admit that individual creative experience and even psychiatric alterations are kinds of knowledge. Certainly the exchanges experienced in reacting to art can be so considered. Admittedly, while realizing various aspects of truth as they emerge during intersubjective exchange, we are inclined to discount or at least limit such a broad view. But since truth is generated through various forms of knowledge, we must at least admit to some aspects of this lenient view.

There is a related issue. Suppose an individual actually knew in detail how his or her brain works. Would we expect that person to abandon his or her reactions to others in terms of propositional attitudes—beliefs, desires, and intentions? I think not. But knowledge of the workings of the brain might at least give that person the ability to reject preposterous assumptions and cant.

At this point, we may further illuminate these issues by setting out a brief summary of the premises of brain-based epistemology. A key question is related to that of A. J. Ayer: How can a system of perceptions be developed and serve as a basis of belief?[8] I would revise this question to add beyond the word "perceptions" the phrase "and an account of consciousness." Ayer has also stated that to know is to be able to perform. This pragmatic statement is good as far as it goes, but it must be modified to include knowledge of moods, for example, where performance is not the issue. Let us see how the premises of brain-based epistemology might lead to useful answers.

First, brain-based epistemology accepts physics and evolutionary biology as essential platforms upon which to base its assertions. It therefore rejects idealist accounts, dualism, panpsychism, and any notion of mental representations proposed without grounding in brain structure. Brain-based epistemology contends that our knowledge is neither a direct copy of our experience nor a direct transfer from our memorial states. Nevertheless, it is entirely compatible with the construction of logical systems on the basis of language and experience, as well as of mathematics as a study of stable mental objects.

It is striking, but perhaps not surprising, that the epistemology of modern physical science is somewhat noncommittal on these issues. By contrast, brain-based epistemology is not noncommittal about the perceiver, who historically and creatively is prior to the science of physics. It is true that we may explain the evolutionary origin of that perceiver by scientific means. But Neural Darwinism and natural selection only provide *grounds* for a series of actual historical and cultural events that affect our knowledge and behavior. Certainly, anyone accepting the tenets of Neural Darwinism and the notions explored here would find it difficult fully to accept the proposals of evolutionary epistemology and the ideas of evolutionary psychology, for both fields are excessively reductive.

We do not inherit a language of thought. Instead, concepts are developed from the brain's mapping of its own perceptual maps. Ultimately, therefore, concepts are initially about the world. Thought itself is based on brain events resulting from the activity of motor regions, activity that does not get conveyed to produce action. It is a premise of brain-based epistemology that subcortical structures such as the basal ganglia are critical in assuring the sequence of such brain events, yielding a kind of presyntax. So thought can occur in the absence of language. In its earliest form, thought is dependent on metaphorical modes and what linguist George Lakoff and philosopher Mark Johnson have called image schemata.[9] This metaphorical activity is strongly buttressed by the associative power of degenerate circuits in the brain. Of course, with the acquisition of language, these powers are enormously expanded. In either case, our brain with its capabilities of pattern recognition, closure, and filling in, goes, as Jerome Bruner pointed out, beyond the information given.[10]

According to brain-based epistemology, the achievement of logic, and to some degree of mathematics, depended on higher-order consciousness, which itself depended for its fullest expression on the acquisition of a true language. The view of brain-based epistemology is that, after the evolution of a bipedal posture, of a supralaryngeal space, of presyntax for movement in the basal ganglia, and of an enlarged cerebral cortex, language arose as an invention. The theory rejects the notion of a brain-based, genetically inherited, language acquisition device. Instead, it contends that language acquisition is epigenetic. Its acquisition and its spread across speech communities would obviously favor its possessors over nonlinguistic hominids even though no direct inheritance of a universal grammar is at issue. Of course, hominids using language could then be further favored by natural selection acting on those systems of learning that favor language skills.

What about the "world" of such linguistically able individuals? What is objective and what is subjective? Brain-based epistemology, having rejected idealism, accepts a position of qualified realism.[11] Its realism is qualified by its recognition of our phenotypic limits. Constraints on our evolved body types, and our brain as a selectional system, clearly allow only a limited sampling of world events, the density of which is enormous. We have already considered the fact that the variation in a selectional brain is to some degree independent of the actual selective events that act to modify the synaptic strengths of particular neuronal groups. There are no infallible or incorrigible mental states in the operation of normal brains. We can even be in error about a phenomenal state—a hallucination can have content but no object. Moreover, we have already discussed the tendency of brain action to find closure, to produce filling in, and to confabulate if necessary. Furthermore, we are possessed of certain necessary illusions. One example is what I have called the Heraclitean illusion—that perceived time is a movement of a period or point progressing from the past through the present to the future. But in fact, the past and the future are concepts; only the remembered present can be linked immediately to actual events in Einsteinian space-time.

Underlying all of these properties is the causal activity of the brain's reentrant thalamocortical system or dynamic core, the complex integrative neural patterns of which entail consciousness. Together with the activity of nonconscious systems, these patterns give rise to learning, memory, and behavior. Behaviorism, philosophical and otherwise, is rejected by brain-based epistemology, which considers that mental acts are conscious. This does not imply that nonconscious brain systems do not have structures and dynamics that interact with and influence the dynamic core. In this respect, Freud's views of unconscious sources of behavior were premonitory.[12] Indeed, the rich interactions of subcortical systems with memory systems in the cortex produce local world events that obviously would never have occurred had conscious organisms such as ourselves not arisen in evolution.

What has emerged from this neural substrate of higher-order consciousness are domains of artistic creation, ethical systems, and a scientific worldview that places us in the order of things. This view provides a source of verifiable truth that enables us to study the brain as the necessary organ for the understanding of all forms of truth. Brain-based epistemology rejects the notion, however, that art, aesthetics, and ethics can be reduced to a series of epigenetic rules of brain action.[13]

The fact that scientific reduction is not exhaustive is no loss. As I have said before, science is imagination in the service of the verifiable truth. Its ultimate power, of course, is in understanding, and as we see around us, its reach in technology is stunning. But the brain origins of imagination in science do not differ from those necessary for poetry, music, or the building of ethical systems. On the model of Neural Darwinism, which recognizes the historical and creative dimensions of human thought, no divorce is necessary between science and the humanities.

Science derives from a variety of cultural events and it generally does not necessarily impel or predict such events. But although scientific theory is necessarily underdetermined, it is as good as we can get. It provides us with the structural conditions of the world's being and of our being, and of how we know them. We can confidently expect that its latest excursion into the analysis of consciousness will further reveal the origins and limits of our second nature, even as it expands and transforms our vision of human knowledge.

Notes

INTRODUCTION

1. H. Adams, *The Education of Henry Adams,* chap. 25, 379.
2. Quine, *Ontological Relativity and Other Essays,* chap. 3.
3. Quine, *Quiddities,* 132–133.
4. James, "Does Consciousness Exist?"

one
THE GALILEAN ARC AND DARWIN'S PROGRAM

1. Whitehead, *Science and the Modern World,* 2.
2. Darwin, *On the Origin of Species.*
3. An excellent account is given in Mayr, *Growth of Biological Thought.*
4. Descartes, "Discourse on the Method" and "Meditations on First Philosophy."
5. Schrödinger, *Mind and Matter.*
6. For a good view of all of these matters, see Heil, *Philosophy of Mind.*
7. Alfred Wallace, the codiscoverer of the theory of natural selection, wrote to Darwin in 1869 expressing what to Darwin was a heretical view. Wallace asserted that the mind and brain of man could not have arisen by natural selection, arguing from the fact that the brains of savages were almost as large as those of Englishmen even though savages were incapable of abstract

thought. Darwin wrote back, saying, "I hope that you have not murdered too completely your child and my own." See Kottler, "Charles Darwin and Alfred Russel Wallace." For a fine account of the background, see Richards, *Darwin and the Emergence of Evolutionary Theories of Mind and Behavior.*

8. Edelman, *Bright Air, Brilliant Fire,* 188.

two

CONSCIOUSNESS, BODY, AND BRAIN

1. Edelman, *Neural Darwinism.* See also my article "Neural Darwinism" in *Neuron.* Both are extensions of the theory first put forth in Edelman and Mountcastle, *The Mindful Brain.* All of these are heavy-going, scholarly works describing a global theory of brain function that has received increasing support over the years. See the books on consciousness below for an abbreviated account.

2. For a brief account, see Edelman, *Wider Than the Sky.* More extensive discussions may be found in Edelman, *The Remembered Present,* and Edelman and Tononi, *A Universe of Consciousness.*

3. For an early account, see Reeke and Edelman, "Real Brains and Artificial Intelligence." See also Searle, *Minds, Brains, and Science.* A further discussion may be found in my book *Bright Air, Brilliant Fire,* 211.

three

SELECTIONISM

1. Mayr, *The Growth of Biological Thought,* gives an excellent discussion of population thinking.

2. A brief account of the immune system and the theory of clonal

selection is given in Edelman, *Bright Air, Brilliant Fire,* chap. 8, "The Sciences of Recognition."

3. Edelman, *Neural Darwinism;* Edelman, *Wider Than the Sky.*

4. This concept is the most challenging one in the theory of neuronal group selection. A simplified account can be found in Edelman, *Wider Than the Sky.*

5. A scholarly review of the dopamine reward system is Unglass, "Dopamine."

6. See Edelman, *Neural Darwinism;* Edelman, *Bright Air, Brilliant Fire;* and Edelman, *The Remembered Present.*

7. This biologically important concept is discussed in my various books. A brief but intensive account is given in Edelman and Gally, "Degeneracy and Complexity in Biological Systems."

four

FROM BRAIN ACTIVITY TO CONSCIOUSNESS

1. This extension of Neural Darwinism or the TNGS is covered in detail in the three books already cited: Edelman, *The Remembered Present;* Edelman, *Wider Than the Sky;* and Edelman and Tononi, *A Universe of Consciousness.*

2. This important point, that qualia are just the discriminations that the dynamic core entails, allowing enhanced adaptive capabilities, is summarized succinctly in Edelman, "Naturalizing Consciousness."

3. This issue is discussed in my article "Naturalizing Consciousness" but also in chap. 7 of my book *Wider Than the Sky;* for a brief history of scientific approaches to consciousness, see Dalton and Baars, "Consciousness Regained."

4. Freud, *On Dreams.*

EPISTEMOLOGY AND ITS DISCONTENTS

1. See, e.g., Dancy and Sosa, eds., *A Companion to Epistemology.*

2. Wittgenstein, *Philosophical Investigations.*

3. Plato argues that we do have knowledge and if this is so, there must be imperceptible Forms which knowledge is about. The Forms are the only real things. Objects of our perception are copies of the Forms and less real. In his dialogue, *Meno,* he claims that an unlettered slave knew the Pythagorean Theorem even before Socrates coached him. This nativism is not unrelated to essentialism. See Plato, "Meno."

4. Descartes, *The Philosophical Writings of Descartes,* trans. Cottingham, Stoothoff, and Murdoch.

5. Rorty, *Philosophy and the Mirror of Nature;* Taylor, "Overcoming Epistemology," 1. A more tightly reasoned account deploring the obsession with epistemology may be found in Searle, "The Future of Philosophy."

6. Quine, "Epistemology Naturalized," in *Ontological Relativity and Other Essays,* 69–90.

7. Piaget, *Genetic Epistemology.* See also Piaget, *Biology and Knowledge.* The examples I give, starting with Quine, are intended to be exemplary, not exhaustive. For a comprehensive account of the various forms of naturalism, see Kitcher, "The Naturalists Return."

8. See, e.g., Messerly, *Piaget's Conception of Evolution.* This book outlines but does not effectively criticize Piaget's "biology."

9. Bishop and Trout, *Epistemology and the Psychology of Human Judgment.*

10. See Campbell, "Evolutionary Epistemology." A comprehensive account may be found in Callebaut and Pinxten, eds., *Evolutionary Epistemology.*

11. Dawkins, *The Selfish Gene.*
12. See Cosmides and Tooby, "From Evolution to Behavior."
13. Lewontin, "Sociobiology—A Caricature of Darwinism"; Gould, *The Mismeasure of Man;* Caplan, ed., *The Sociobiology Debate.*

six

A BRAIN-BASED APPROACH

1. Boyd and Richerson, *The Origin and Evolution of Cultures.*
2. Merzenich, Nelson, Stryker, Schoppman, and Zook, "Somato-sensory Cortical Map Changes Following Digit Manipulation in Adult Monkeys."
3. For a rich analysis of metaphor in bodily terms, see Lakoff, *Women, Fire, and Dangerous Things.* Another related account is Johnson, *The Body in the Mind.* A synthesis with an emphasis on psychology comes from a well-informed psychiatrist: Modell, *Imagination and the Meaningful Brain.*
4. The outstanding figure here is Noam Chomsky. See his classic account in *Cartesian Linguistics.* His more recent thoughts are embodied in the books *Some Concepts and Consequences of the Theory of Government and Binding* and *Language and Thought.*
5. Tarski, "The Concept of Truth in Formalized Languages."
6. This whole discussion is not free from controversy. See, e.g., the contrasting views in two recent publications and a perspective on both: Lemer, Izard, and Dehaene, "Exact and Approximate Arithmetic in an Amazonian Indigene Group"; Gordon, "Numerical Cognition without Words"; and the perspective on these reports, Gelman and Gallistiel, "Language and the Origin of Numerical Concepts."

7. Carey, "Bootstrapping and the Origin of Concepts." For a general background, see Dehaene, *The Number Sense*.

8. Another quotation, presumably translated, is: "God created the integers, all else is the work of man." See Bell, *Men of Mathematics*, 477.

9. Edelman and Gally, "Degeneracy and Complexity in Biological Systems."

10. The idea derives from Hume. The naturalistic fallacy was pointed out in Moore, *Principia Ethica*.

11. I use the term "second nature" to refer to the sum of our experienced perceptions, memories, and attitudes individually and collectively. The term is perhaps best encapsulated in the notion of common sense knowledge derived from everyday experience rather than from scientific knowledge. This usage should not be conflated with the distinction between the Manifest Image and Scientific Image drawn by the philosopher Wilfred Sellars. As he puts it, the Manifest Image is the commonsense framework of man-in-the-world, but it also includes correlational and inductive science. The Scientific Image embodies the postulated entities of theoretical science, for example, atoms, molecules, and microphysics. Thus, both images invoke scientific knowledge. Sellars's distinctions were aimed at philosophers. My usage is more modest and is simply intended to contrast our everyday impressions and conclusions with those reached through scientific pursuits. See Sellars, "Philosophy and the Scientific Image of Man." For the kind of contrast between second nature and nature that I have in mind, see Eddington, *The Nature of the Physical World,* ix–xii. This gifted astronomer contrasts the table before him—"strange compound of external nature, mental imagery, and inherited prejudice"—with the scientific description of his table, "mostly emptiness full of speedy electric charges."

12. Boyd and Richerson, *The Origin and Evolution of Cultures*.
13. Huxley, "On the Method of Zadig." In this essay, based on a talk, Huxley points out that "prophecy" is not necessarily a fore-telling of the future but rather, as Voltaire points out in his fantasy, "Zadig," it can consist of insights derived from present evidence concerning events in the past.

seven

FORMS OF KNOWLEDGE

1. G. Sarton, *Appreciation of Ancient and Medieval Science during the Renaissance*.
2. Vico, *The New Science of Giambattista Vico;* Berlin, *Vico and Herder*.
3. Berlin, "The Divorce between the Sciences and the Humanities," 326.
4. Dilthey's *Philosophy of Existence*.
5. Vico, *The New Science of Giambattista Vico;* Berlin, *Vico and Herder: Two Studies in the History of Ideas*.
6. James, "Does Consciousness Exist?"
7. Whitehead, *Modes of Thought*.
8. Snow, *The Two Cultures and Scientific Revolution*.
9. Schrödinger, *Mind and Matter*.
10. Watson, *Behaviorism;* Skinner, *About Behaviorism*.
11. Churchland, *The Engine of Reason*.
12. Otto Neurath was a critical figure in the so-called Vienna Circle and in later life sponsored the Unity of Science Movement and published the *Encyclopedia of Unified Science*. See "Sociology and Physicalism, Erkenntnis 2 (1931–2)" and "Protocol Sentences (1932–3)" in Ayer, ed. *Logical Positivism*.

13. Weinberg, *Dreams of a Final Theory*. Against the notion of a TOE we have Laughlin and Pines, "The Theory of Everything." Laughlin has written an extensive account against extreme reductionism in *A Different Universe*.

14. Wilson, *Consilience*. Stephen Jay Gould wrote an impassioned critique of Wilson's position in *The Hedgehog, the Fox, and the Magister's Pox*. See especially chapter 9, "The False Path of Reductionism and the Consilience of Equal Regard."

15. Wilson, *Consilience*, 11.

eight
REPAIRING THE RIFT

1. Berlin, "The Concept of Scientific History."

2. Hempel, *Aspects of Scientific Explanation and Other Essays in the Philosophy of Science*.

3. These, in classical terms, are propositional attitudes, states of mind having propositional contents and attitudes toward them. They include beliefs, desires, intentions, wishes, fears, doubts, and hopes.

4. B. Adams, *The Law of Civilization and Decay*.

5. Spengler, *The Decline of the West;* Toynbee, *A Study of History*. These two and Adams may be looked on as metahistorians or bold synthesizers, admirable for their sweep if not for their judgments.

6. Gaddis, *The Landscape of History*.

7. See Edelman, *Wider Than the Sky*, 147–148.

8. Lakoff, *Women, Fire, and Dangerous Things*.

9. Wilson, *Consilience;* Gould, *The Hedgehog, the Fox, and the Magister's Pox*.

10. D. A. Hume, *Treatise of Human Nature;* Moore, *Principia Ethica.*
11. The philosopher Avrum Stroll has argued strongly that there are questions of fact that "even in principle" science cannot answer. See Stroll, *Did My Genes Make Me Do It?*
12. Quine, *Ontological Relativity and Other Essays;* Edelman, *Wider Than the Sky.*

nine

CAUSATION, ILLUSIONS, AND VALUES

1. Van't Hoff, *Imagination in Science.*
2. Intentionality is considered extensively in Searle, *Consciousness and Language.*
3. Quine, *Word and Object.*
4. Epiphenomenalism is sometimes seen as a cousin of dualism, as an objectionable spooky doctrine. But the color (or more rightly, the spectrum) of hemoglobin which is entailed by the molecule's structure requires no such doctrine. The spectrum is not causal, but the color changes when, causally, oxygen is bound.
5. For an account suggesting that conscious will has many illusory properties, see Wegener, *The Illusion of Conscious Will.*
6. Damasio, *The Feeling of What Happens.*

ten

CREATIVITY

1. Edelman, *Bright Air, Brilliant Fire,* chap. 8.
2. This quotation has been attributed to a character in E. M. Forster's

novel *Howard's End* (1910), though I have not been able to locate it there. Nonetheless, it has been attributed to him.

3. This view of thought as being essentially motoric is consistent with the known interactions of the prefrontal and parietal cortex with basal ganglia, the subcortical regions involved in motor programs; see figure 1. The essential notion is that all that is motoric is not necessarily expressed as action or movement.

4. Kanizsa, *Organization in Vision.*

eleven

ABNORMAL STATES

1. Even so, there are numerous subtleties in diagnosing various syndromes and states of neural illness. The nature of the task becomes obvious on consulting the *Diagnostic and Statistical Manual of Mental Disorders: DSM-IV-TR.*

2. Freud, *Standard Edition.*

3. For an account of how Haeckel's biogenetic law fell into disregard, see S. J. Gould, *Ontogeny and Phylogeny.*

4. Two relatively nontechnical books on neuropsychological syndromes are Feinberg, *Altered Egos,* and Hirstein, *Brain Fiction.* Hirstein focuses on syndromes leading to confabulation, a subject that obviously has a bearing on how we know and know we know. The neurologist Oliver Sacks has written insightfully about the effects of neuropsychological syndromes on modes of being and knowing. His accounts beautifully describe the ways in which alterations in the nervous system are reflected in behavior. See *The Man Who Mistook His Wife for a Hat.*

5. Sperry, "Some Effects of Disconnecting the Cerebral Hemispheres."

6. See Feinberg, *Altered Egos,* and Hirstein, *Brain Fiction.*

7. See Hirstein, *Brain Fiction.*

8. This no place for textbook references on either psychosis or neurosis. A perusal of apposite sections in *DSM–IV-TR* will give sufficient detail.

9. See Wollheim, *Freud.*

10. See the quotation in Curtis and Greenslet, eds., *The Practical Cogitator,* 31–35.

twelve

BRAIN-BASED DEVICES

1. Krichmar and Edelman, "Brain-Based Devices for the Study of Nervous Systems and the Development of Intelligent Machines."

2. Turing, "On Computable Numbers, with an Application to the *Entscheidungs* Problem."

3. Hunt, *Understanding Robotics,* 7.

4. The device Darwin VII is not described in detail in Krichmar and Edelman, "Brain-Based Devices." It resembled Darwin VIII closely but had no reentrant structures in its brain. For details, see Krichmar and Edelman, "Machine Psychology." Later refinements are in Krichmar, Nitz, Gally, and Edelman, "Characterizing Functional Hippocampal Pathways in a Brain-based Device as It Solves a Spatial Memory Task."

5. Such a perception-Turing machine would combine the perceptual and learning abilities of a BBD-like portion with the knowledge base and calculating powers of a digital computer. Mistakes would have to be avoided in the computer portion while the perception "machine" would deal with novelty, which in its na-

ture is not programmable. At the same time the perception machine would learn by making mistakes. Mutual communication between the two portions of such a machine should lead to great increases in computing power and pattern recognition.

thirteen
SECOND NATURE

1. For the Darwin-Wallace correspondence, see Kottler, "Charles Darwin and Alfred Russel Wallace."
2. Quine, *Pursuit of Truth,* 71.
3. Brentano expanded on the notion of intentionality, which he considered to distinguish the mental from the physical. For a modern exposition of the concept, see the collection of essays by Searle, *Consciousness and Language.* In later life, Brentano became an explicit dualist. His key early work was Brentano, *Psychology from an Empirical Standpoint.*
4. Bishop and Trout, *Epistemology and the Psychology of Human Judgment.*
5. Blackburn, *Truth: A Guide;* Lynch, *True to Life.*
6. Changeux, *The Physiology of Truth.* This book uses a version of Neural Darwinism and the theory of reentry to claim that selection in evolution provided a basis for truth—a physiology of truth. But this position fails to recognize that the search for truth is, in Stephen Jay Gould's word, an exaptation. Selection for consciousness may provide adaptive advantage for planning but is no guarantee of truth. The claim, even metaphorical, that there is a physiology of truth is ill-founded. The assumption of Popper's model of how knowledge evolves is also unconvincing, given the evidence of our irrational behavior. One hope for guidance

derived from epistemology comes from a pragmatic view of reasoning as suggested by Bishop and Trout. For strong arguments that our brains have not evolved directly to achieve knowledge of truth, see Kitcher, "The Naturalists Return," and Stich, *The Fragmentation of Reason*.

7. Goldman, *Knowledge in a Social World*. See also Kitcher, *The Advances of Science*.
8. Ayer, *Philosophy in the Twentieth Century*.
9. Lakoff, *Women, Fire, and Dangerous Things*; Johnson, *The Body in the Mind*.
10. Bruner, *Going beyond the Information Given*.
11. Edelman, *The Remembered Present*.
12. Wollheim, *Freud*.
13. Gould, *The Hedgehog, the Fox, and the Magister's Pox*.

Bibliography

Adams, B. *The Law of Civilization and Decay: An Essay on History.* New York: Macmillan, 1896. Reprint ed., New York: Gordon, 1943.

Adams, H. *The Education of Henry Adams.* Boston: Houghton Mifflin, 1973.

Ayer, A. J., ed. *Logical Positivism.* New York: Free Press, 1959.

———. *Philosophy in the Twentieth Century.* East Hanover, NJ: Vintage Books, 1984.

Bell, E. T. *Men of Mathematics: The Lives and Achievements of the Great Mathematicians from Zeno to Poincaré.* New York: Simon and Schuster, 1986.

Berlin, I. "The Concept of Scientific History." In Berlin, *The Proper Study of Mankind: An Anthology of Essays,* 17–58. New York: Farrar, Straus and Giroux, 1997.

———. "The Divorce between the Sciences and the Humanities." In Berlin, *The Proper Study of Mankind,* 320–358. New York: Farrar, Straus, and Giroux, 1997.

———. *Vico and Herder: Two Studies in the History of Ideas.* New York: Viking, 1976.

Bishop, M. A., and J. D. Trout. *Epistemology and the Psychology of Human Judgment.* New York: Oxford University Press, 2005.

Blackburn, S. *Truth: A Guide.* New York: Oxford University Press, 2005.

Boyd, R., and P. J. Richerson. *The Origin and Evolution of Cultures.* New York: Oxford University Press, 2005.

Brentano, F. *Psychology from an Empirical Standpoint.* 2nd ed. Trans. A. C. Rancurello, D. B. Terrell, and L. L. McAlister. London: Routledge, 1995.

Bruner, J. *Going beyond the Information Given.* New York: Norton, 1993.

Callebaut, W., and R. Pinxten, eds. *Evolutionary Epistemology: A Multiparadigm Program.* Synthese Library, 190. Dordrecht: Reidel, 1987.

Campbell, D. T. "Evolutionary Epistemology." In P. A. Schlipp, ed., *The Philosophy of Karl Popper,* 412–463. La Salle, IL: Open Court, 1974.

Caplan, A. L., ed. *The Sociobiology Debate.* New York: Harper and Row, 1978.

Carey, S. "Bootstrapping and the Origin of Concepts." *Daedalus* 133 (2004): 59–68.

Changeux, J.-P. *The Physiology of Truth: Neuroscience and Human Knowledge.* Trans. M. B. DeBevoise. Cambridge, MA: Belknap Press of Harvard University Press, 2004.

Chomsky, N. *Cartesian Linguistics.* New York: Harper and Row, 1966.

———. *Language and Thought.* Wakefield, RI: Moyer Bell, 1993.

———. *Some Concepts and Consequences of the Theory of Government and Binding.* Cambridge, MA: MIT Press, 1982.

Churchland, P. *The Engine of Reason, the Seat of the Soul: Philosophical Journey into the Brain.* Cambridge, MA: MIT Press, 1996.

Cosmides, L., and J. Tooby. "From Evolution to Behavior: Evolutionary Psychology as the Missing Link." In J. Dupré, ed., *The Latest on the Best: Essays on Evolution and Optimality,* 277–306. Cambridge, MA: MIT Press, 1987.

Curtis, C. P., Jr., and F. Greenslet, eds. *The Practical Cogitator; or, The Thinker's Anthology.* Boston: Houghton Mifflin, 1962.

Dalton, T. C., and B. J. Baars. "Consciousness Regained: The Scientific Restoration of Mind and Brain." In Dalton and R. B. Evans, eds., *The Life Cycle of Psychological Ideas,* 203–247. New York: Kluwer Academic/Plenum, 2004.

Damasio, A. R. *The Feeling of What Happens.* New York: Harcourt Brace, 1999.

Dancy, J., and E. Sosa, eds. *A Companion to Epistemology.* Oxford: Blackwell, 1992.

Darwin, C. *On the Origin of Species by Means of Natural Selection, or the Preservation of Favored Races in the Struggle for Life.* London: John Murray, 1859.

Dawkins, R. *The Selfish Gene.* New York: Oxford University Press, 1976.

Dehaene, S. *The Number Sense.* Oxford: Oxford University Press, 1997.

Descartes, R. "Discourse on the Method." In *The Philosophical Writings of Descartes,* trans. J. Cottingham, R. Stoothoff, and D. Murdoch, vol. 1, 109–176. Cambridge: Cambridge University Press, 1984.

———. "Meditations on First Philosophy." In *The Philosophical Writings of Descartes,* trans. J. Cottingham, R. Stoothoff, and D. Murdoch, vol. 2, 1–49. Cambridge: Cambridge University Press, 1984.

Diagnostic and Statistical Manual of Mental Disorders: DSM-IV-TR. 4th ed., text revision. Washington, DC: American Psychiatric Association, 2000.

Dilthey, Wilhelm. *Philosophy of Existence: Introduction to Weltanschauungslehre.* Trans. W. Kluback and M. Weinbaum. New York: Bookman, 1957.

Eddington, A. S. *The Nature of the Physical World.* Cambridge: Cambridge University Press, 1929.

Edelman, G. M. *Bright Air, Brilliant Fire: On the Matter of the Mind.* New York: Basic Books, 1992.

———. "Naturalizing Consciousness: A Theoretical Framework." *Proceedings of the National Academy of Sciences USA* 100 (2003): 5520–5524.

———. *Neural Darwinism: The Theory of Neuronal Group Selection.* New York: Basic Books, 1987.

———. *The Remembered Present: A Biological Theory of Consciousness.* New York: Basic Books, 1989.

———. *Wider Than the Sky: The Phenomenal Gift of Consciousness.* New Haven and London: Yale University Press, 2004.

———, and J. A. Gally. "Degeneracy and Complexity in Biological Systems." *Proceedings of the National Academy of Sciences USA* 98 (2001): 13763–13768.

———, and V. B. Mountcastle. *The Mindful Brain: Cortical Organization and the Group-Selective Theory of Higher Brain Function.* Cambridge, MA: MIT Press, 1978.

———, and G. Tononi. *A Universe of Consciousness: How Matter Becomes Imagination.* New York: Basic Books, 2000.

Feinberg, T. E. *Altered Egos: How the Brain Creates the Self.* New York: Oxford University Press, 2001.

Freud, S. *On Dreams.* Ed. J. Strachey. Reprint ed., New York: Norton, 1963.

———. *The Standard Edition of the Complete Psychological Works of Sigmund Freud.* 24 vols. Trans. J. Strachey in collaboration with A. Freud, assisted by A. Strachey and A. Tyson. London: Hogarth Press and Institute of Psychoanalysis, 1975.

Gaddis, J. L. *The Landscape of History: How Historians Map the Past.* New York: Oxford University Press, 2002.

Gelman, R., and C. R. Gallistiel. "Language and the Origin of Numerical Concepts." *Science* 306 (2004): 441–443.

Goldman, A. I. *Knowledge in a Social World.* Oxford: Clarendon Press, 1999.

Gordon, P. "Numerical Cognition without Words: Evidence from Amazonia." *Science* 306 (2004): 496–499.

Gould, S. J. *The Hedgehog, The Fox, and the Magister's Pox: Minding the Gap between Science and the Humanities.* New York: Harmony Books, 2003.

———. *The Mismeasure of Man.* New York: W. W. Norton, 1981.

———. *Ontogeny and Phylogeny.* Cambridge, MA: Belknap Press of Harvard University Press, 1977.

Heil, J. *Philosophy of Mind: A Guide and Anthology.* Oxford: Oxford University Press, 2004.

Hempel, C. G. *Aspects of Scientific Explanation and Other Essays in the Philosophy of Science.* New York: Free Press, 1965.

Hirstein, W. *Brain Fiction: Self-Deception and the Riddle of Confabulation.* Cambridge, MA: MIT Press, 2005.

Hume, D. *A Treatise of Human Nature.* London: Routledge and Kegan Paul, 1985.

Hunt, V. D. *Understanding Robotics.* New York: Academic Press, Harcourt Brace Jovanovich, 1990.

Huxley, T. H. "On the Method of Zadig: Retrospective Prophecy as a Function of Science." In *Science and Hebrew Tradition: Essays by Thomas H. Huxley,* 1–22. New York: D. Appleton, 1894.

James, W. "Does Consciousness Exist?" In James, *Essays in Radical Empiricism,* 1–38. New York: Longman Green, 1912.

Johnson, M. *The Body in the Mind.* Chicago: University of Chicago Press, 1987.

Kanizsa, G. *Organization in Vision.* New York: Praeger, 1979.

Kitcher, P. *The Advances of Science.* New York: Oxford University Press, 1993.

———. "The Naturalists Return." *Philosophical Review* 101, no. 1 (1992): 53–114.

Kottler, M. J. "Charles Darwin and Alfred Russel Wallace: Two Decades of Debate over Natural Selection." In D. Kohn, ed., *The Darwinian Heritage,* 367–432. Princeton, NJ: Princeton University Press, 1985.

Krichmar, J. L., and G. M. Edelman. "Brain-Based Devices for the Study of Nervous Systems and the Development of Intelligent Machines." *Artificial Life* 111 (2005): 67–77.

———. "Machine Psychology: Autonomous Behavior, Perceptual Categorization and Conditioning in a Brain-based Device." *Cerebral Cortex* 12 (2002): 818–830.

Krichmar, J. L., D. A. Nitz, J. A. Gally, and G. M. Edelman. "Characterizing Functional Hippocampal Pathways in a Brain-based Device as It Solves a Spatial Memory Task." *Proceedings of the National Academy of Sciences USA* 102 (2005): 2111–2116.

Lakoff, G. *Women, Fire, and Dangerous Things.* Chicago: University of Chicago Press, 1987.

Laughlin, R. B. *A Different Universe: Reinventing Physics from the Bottom Down.* New York: Basic Books, 2005.

———, and D. Pines. "The Theory of Everything." *Proceedings of the National Academy of Science USA* 97 (2000): 28–31.

Lemer, C., V. Izard, and S. Dehaene. "Exact and Approximate Arithmetic in an Amazonian Indigene Group." *Science* 306 (2004): 499–503.

Lewontin, R. "Sociobiology—A Caricature of Darwinism." In P. Asquith and F. Suppe, eds., *PSA 1976,* 2:22–31. East Lansing, MI: Philosophy of Science Association 1977.

Lynch, M. P. *True to Life: Why Truth Matters.* Cambridge, MA: MIT Press, 2004.

Mayr, E. *The Growth of Biological Thought: Diversity, Evolution, and Inheritance.* Cambridge, MA: Harvard University Press, 1982.

Merzenich, M. M., R. J. Nelson, M. P. Stryker, A. Schoppman, and J. M. Zook. "Somatosensory Cortical Map Changes Following Digit Manipulation in Adult Monkeys." *Journal of Comparative Neurology* 224 (1984): 591–605.

Messerly, J. G. *Piaget's Conception of Evolution: Beyond Darwin and Lamarck.* Lanham, MD: Bowman and Littlefield, 1996.

Modell, A. H. *Imagination and the Meaningful Brain.* Cambridge, MA: MIT Press, 2003.

Moore, G. E. *Principia Ethica.* Cambridge: Cambridge University Press, 1903.

Piaget, J. *Biology and Knowledge: An Essay on the Relations between Organic Regulations and Cognitive Processes.* Chicago: University of Chicago Press, 1971.

———. *Genetic Epistemology.* New York: Columbia University Press, 1970.

Plato. "Meno." In E. Hamilton and H. Cairns, eds., *The Collected Dialogues of Plato.* Princeton, NJ: Princeton University Press, 1961.

Quine, W. V. *Ontological Relativity and Other Essays.* New York: Columbia University Press, 1969.

———. *Pursuit of Truth.* Cambridge, MA: Harvard University Press, 1990.

———. *Quiddities: An Intermittently Philosophical Dictionary.* Cambridge, MA: Belknap Press of Harvard University Press, 1987.

———. *Word and Object.* Cambridge, MA: MIT Press, 1960.

Reeke, G. N., Jr., and G. M. Edelman. "Real Brains and Artificial Intelligence." *Daedalus* 117 (1987): 143–173.

Richards, R. J. *Darwin and the Emergence of Evolutionary Theories of Mind and Behavior.* Chicago: University of Chicago Press, 1987.

Rorty, R. *Philosophy and the Mirror of Nature.* Princeton, NJ: Princeton University Press, 1979.

Sacks, O. *The Man Who Mistook His Wife for a Hat and Other Clinical Tales.* New York: Simon and Schuster, 1998.

Sarton, G. *Appreciation of Ancient and Medieval Science during the Renaissance.* New York: Barnes, 1955.

Schrödinger, E. *Mind and Matter.* Cambridge: Cambridge University Press, 1958.

Searle, J. R. *Consciousness and Language.* Cambridge: Cambridge University Press, 2002.

———. "The Future of Philosophy." *Philosophical Transactions of the Royal Society London, B.* 354 (1999): 2069–2080.

———. *Minds, Brains and Science.* Cambridge, MA: Harvard University Press, 1984.

Sellars, W. "Philosophy and the Scientific Image of Man." In Sellars, *Science, Perception and Reality,* 1–40. London: Routledge and K. Paul, 1963.

Skinner, B. F. *About Behaviorism.* New York: Vintage, 1976.

Snow, C. P. *The Two Cultures and Scientific Revolution.* New York: Norton, 1930.

Spengler, O. *The Decline of the West.* New York: Alfred Knopf, 1939.

Sperry, R. W. "Some Effects of Disconnecting the Cerebral Hemispheres." Nobel lecture. *Les Prix Nobel.* Stockholm: Almqvist & Wiksell, 1981.

Stich, S. *The Fragmentation of Reason.* Cambridge, MA: MIT Press, 1990.

Stroll A. *Did My Genes Make Me Do It? And Other Philosophical Dilemmas.* Oxford: One World, 2004.

Tarski, A. "The Concept of Truth in Formalized Languages." In Tarski, *Logic, Semantics, Metamathematics: Papers from 1923 to 1938,* 152–278. Trans. J. H. Woodger. Oxford: Clarendon Press, 1956.

Taylor, C. "Overcoming Epistemology." In Taylor, *Philosophical Arguments,* 1–19. Cambridge, MA: Harvard University Press, 1995.

Toynbee, A. *A Study of History.* Abridgement by D. C. Somerveld. 2 vols. Oxford: Oxford University Press, 1957.

Turing, A. "On Computable Numbers, with an Application to the *Entscheidungs* Problem." *Proceedings of the London Mathematical Society,* 2nd Ser., 42 (1936): 230–265.

Unglass, M.A. "Dopamine: The Salient Issue." *Trends in Neurosciences* 27 (2004): 702–706.

van't Hoff, J. H. *Imagination in Science.* Trans. G. F. Springer. Berlin: Springer-Verlag, 1967.

Vico, G. B. *The New Science of Giambattista Vico* (1744). Trans. T. G. Bergin and M. H. Fisch. Ithaca, NY: Cornell University Press, 1948; reprint ed., Cornell Paperback, 1976.

Watson, J. *Behaviorism.* New York: Norton, 1930.

Wegener, D. M. *The Illusion of Conscious Will.* Cambridge, MA: MIT Press, 2003.

Weinberg, S. *Dreams of a Final Theory: The Scientist's Search for the Ultimate Laws of Nature.* New York: Vintage, 1994.

Whitehead, A. N. *Modes of Thought.* New York: Macmillan, 1938.

———. *Science and the Modern World.* New York: Macmillan, 1925. Reprint ed., New York: Free Press, 1967.

Wilson, E. O. *Consilience: The Unity of Human Knowledge.* New York: Vintage, 1999.

Wittgenstein, L. *Philosophical Investigations.* 3rd ed. New York: Macmillan, 1953.

Wollheim, Richard. *Freud: A Collection of Critical Essays.* Garden City, NY: Anchor Press/Doubleday, 1974.

Index

Italic page numbers refer to figures

antibodies, 26–27, 100–101, 102

antigens, 26–27, 100, 101, 102

Anton's syndrome, 114

anxiety disorders, 118

apraxias, 110

artificial intelligence, 21

assimilation, 48

association: and memory, 33, 34, 36; and degenerate circuits in brain, 34, 83, 153; and reentrant degenerate system, 58; and metaphor, 61, 83, 122; and discriminations, 90; and reentrant interactions in brain, 151

attention, 39, 86, 94, 102, 115

Australopithicus, 55

Ayer, A. J., 152

Bacon, Francis, 70, 82

basal ganglia: and brain structure, *16,* 17; and signal pathways, 20; and value systems, 59; and language, 61, 154; and brain-based devices, 139; and sequence of brain events, 153; prefrontal and parietal cortex interactions with, 168n3

basal syntax, 61

behavior: and animals, 15, 128–129; and computer metaphor for brain, 21; and brain as selectional system, 28; and value systems, 31; and nonconscious interactions, 39; neural constraints on, 50; and evolutionary psychology, 51–52; and perceptual categorization, 59; adaptive behavior, 65; rational behavior, 81; regularities of, 82, 84; effects of unconscious processes on, 108; and neuroses, 122; and brain-based devices, 139; historical and cultural events affecting, 153; and brain-based epistemology, 155

behaviorism, 73, 155

beliefs: meaning of, 44; and Descartes, 45; psychological processes leading to, 46; and thought, 63, 90, 123; cultural factors in, 66; and history, 78; reflected in consciousness, 89; origins of, 109, 148; and brain-body relationship, 114; and psychoanalytic theory, 121; and perceptions, 152

Berlin, Isaiah, 70, 77, 78–79, 80, 81

biology: and conception of nature, 6; of Piaget, 48–49; and conscious thought, 91; and physics, 149

brain as selectional system
(continued)
130; and creativity, 97, 99,
104; and perceptual categorization, 101; and conscious
artifacts, 126; and brain-
based devices, 131, 134; and
individual activity patterns,
135–136; and consciousness,
143, 170n6; and sampling of
world events, 154–155. *See
also* Neural Darwinism
brain-based devices (BBDs), 127,
128, 130–131, *132,* 133–141,
169–170n5
brain-based epistemology: and
brain function, 2; grounds for,
42, 52; and knowledge acquisition, 54, 66; and natural selection, 54, 64; and discriminatory responses, 60; and
language, 60–62, 65–66, 154;
and heterogeneous sources of
knowledge, 64–65; and traditional epistemology, 65, 104;
recursive element in, 84–85;
naturalized epistemology
contrasted with, 86, 147; and
perceptions, 86, 152; and
causality, 91; and generativity,
95–96; and Freud's psychoanalytic theory, 120; and ab-

normal states, 124; and data
from neuroscience and psychology, 148; and truth, 148,
156; and normative issues,
149; premises of, 152–156;
and qualified realism, 154–155
brain events: and scientific
methodology, 7–8; unique-
ness of, 84; recombinatorial
and integrative power of, 100;
scientific observation of, 146
brain function, 2, 65, 77, 108
brain maps, 56–57
brain stem, *16,* 17, 30, 59
brain theory, 8, 10, 14, 69
Brentano, Franz, 147–148, 170n3
Broca's aphasia, 109–110
Bruner, Jerome, 154

C, 92, 145
C′, 92, 145
Campbell, Donald, 51
Capgras's syndrome, 114
Cartesian observer, 55
causality: and notion of zombie,
40–41; physical causation,
40, 64, 71, 74; and thalamocortical system, 40, 92; and
epistemological issues, 47,
91; and study of consciousness, 89; and macroscopic
level of physical world,

Index

science (continued)
    and science wars, 71; and
    logical positivism, 73; histori-
    cal context of, 82, 85, 87; aim
    of, 89; and error, 105. *See also*
    neuroscience
science and humanities: differ-
    ing views of world, 1–2;
    divorce between, 2, 8, 10,
    69–75, 77, 84; relation
    between, 85–87, 91, 156
scientific descriptions, experience
    distinguished from, 150, 151
scientific explanations, and
    everyday experience, 2, 66,
    164n11
scientific history, 77–81
Scientific Image, 164n11
scientific inquiry: and conscious-
    ness, 8, 9, 149, 156–157; and
    verifiable truth, 8, 66, 69, 71,
    85, 89, 105, 156; and forma-
    tion of precise concepts, 60, 84
scientific insight, 91
scientific methodology, and
    brain events, 7–8
scientific reductionism: and
    theory of everything, 73–74;
    and science and humanities,
    77, 85–86; and history, 79;
    and consilience, 84; limita-
    tions on, 91; and theory of

neuroses, 109; and epigenetic
    rules of brain science, 143;
    and ethics and aesthetics,
    146; and evolutionary episte-
    mology, 153
scientific technology, 6
second nature: meaning of, 4,
    164n11; conception of, 6;
    scientific explanation of, 66,
    146; observation of outside
    world contrasted with, 70;
    and creativity, 99–100; recon-
    ciliation with nature, 143; and
    truth, 150–151; origin and
    limits of, 157. *See also* nature
selectional systems: immune
    system as, 26–27, 100–101,
    102; and degeneracy, 33–34,
    83; and generation of diver-
    sity, 100; and specificity and
    range, 102, 103, 122; and
    brain-based devices, 130.
    See also brain as selectional
    system
selectionism, and knowledge,
    51–52
self: nameable self, 15; process
    of emergence of, 37–38;
    social self, 38, 61; reliance
    on conscious experience, 83;
    cognitive and emotional con-
    struction of, 95; and Freud's

synaptic connections: between neurons, *16,* 17, 18, 55; recording of, 18–19; and correlation of firing patterns, 28; and brain-based devices, 134

syntax, 38, 60–61, 110, 139

Tarski, Alfred, 62

Taylor, Charles, 45

tertiary syphilis, 116

thalamocortical system: diagram of, *29;* and primary consciousness, 36; and qualia, 37; evolution of, 39, 58; and causality, 40, 92; and neuropsychological syndromes/diseases, 115–116; and brain-based devices, 127, 139; and neural correlates of consciousness, 144

thalamus: and brain structure, *16,* 17; and signal pathways, 19, 20; mediodorsal nucleus of, 115. *See also* thalamocortical system

theory of everything (TOE), 73–74

theory of neuronal group selection (TNGS), 30, 54, 57, 60, 61, 64. *See also* Neural Darwinism

thought: related to emotion, 10, 90; and brain speaking to itself, 57; and metaphor,

58–59, 65, 90, 122, 153; early formulations of, 60; preceding language, 63, 153; biological origins of truth, 69, 96; neural basis of, 90, 91, 153; and combinatorial freedom, 103; logic as mode of, 103; as motoric without action, 103, 123, 153, 168n3; pattern recognition as mode of, 103; inappropriate thoughts, 115; and psychoanalytic theory, 121; and sensations, 123

time, passage of, 93

touch, 56

Toynbee, Arnold, 79, 166n5

traditional epistemology: and knowledge as justified true belief, 44–46, 96; and language, 44, 60, 63–64, 96; and validity and justification, 48, 50; critics of, 49, 55; and development of knowledge, 64; and brain-based epistemology, 65, 104; scientific basis for, 67; domains excluded from, 148

Trout, J. D., 49–50, 171n6

true language, 38, 90, 139, 154

truth: search for, 4; science in service of verifiable truth, 8, 66, 69, 71, 85, 89, 105, 156; and knowledge, 10, 151;